BestMasters

Mit „**BestMasters**" zeichnet Springer die besten Masterarbeiten aus, die an renommierten Hochschulen in Deutschland, Österreich und der Schweiz entstanden sind. Die mit Höchstnote ausgezeichneten Arbeiten wurden durch Gutachter zur Veröffentlichung empfohlen und behandeln aktuelle Themen aus unterschiedlichen Fachgebieten der Naturwissenschaften, Psychologie, Technik und Wirtschaftswissenschaften. Die Reihe wendet sich an Praktiker und Wissenschaftler gleichermaßen und soll insbesondere auch Nachwuchswissenschaftlern Orientierung geben.

Springer awards "**BestMasters**" to the best master's theses which have been completed at renowned Universities in Germany, Austria, and Switzerland. The studies received highest marks and were recommended for publication by supervisors. They address current issues from various fields of research in natural sciences, psychology, technology, and economics. The series addresses practitioners as well as scientists and, in particular, offers guidance for early stage researchers.

Oona Rössler

SARP-Driven Activation of Antibiotic Biosynthetic Gene Clusters in Actinomycetes

 Springer Spektrum

Oona Rössler
Department of Bioresources for
Bioeconomy and Health Research
Leibniz Institute DSMZ – German
Collection of Microorganisms and Cell
Cultures
Braunschweig, Germany

Dissertation an der Technischen Universität Braunschweig, Fakultät für Lebenswissenschaften.

ISSN 2625-3577 ISSN 2625-3615 (electronic)
BestMasters
ISBN 978-3-658-44551-5 ISBN 978-3-658-44552-2 (eBook)
https://doi.org/10.1007/978-3-658-44552-2

This Springer Spektrum imprint is published by the registered company Springer Fachmedien Wiesbaden GmbH, part of Springer Nature.
The registered company address is: Abraham-Lincoln-Str. 46, 65189 Wiesbaden, Germany

Paper in this product is recyclable.

Acknowledgments

I would like to thank my first examiner and supervisor Prof. Dr. Yvonne Mast (Leibniz-Institute DSMZ) for the opportunity to perform my master project in her research group and for her great support during and after the time. In addition, I would like to thank my second examiner Prof. Dr. Dieter Jahn (Technical University Braunschweig). I would like to express special thanks to Dr. Juan Pablo Gomez-Escribano (Leibniz-Institute DSMZ) for his support, advice, encouragement, and fun during his supervision. Additionally, I gratefully acknowledge the experimental support of Dr. Anja Schüffler (IBWF), Prof. Dr. Heike Brötz-Oesterheldt (University of Tübingen) und Anne Wochele (University of Tübingen). I also would like to thank Prof. Dr. Blankenfeldt und Dr. Konrad Büssow (HZI) for providing the purified PapR2 proteins. Finally, I must express my profound gratitude to my parents and my boyfriend Michele for their unconditional support and continuous encouragement throughout my years of study.

Abstract

Actinomycetes are a group of Gram-positive bacteria of which many representatives are prominent for being prolific producers of bioactive natural products including antibiotics, fungicides, antitumor agents, or immunosuppressants. *Streptomyces* antibiotic regulatory protein (SARP) family of transcriptional regulators are widely distributed among actinomycetes, especially in streptomycetes and are known to activate antibiotic biosynthesis. The set of genes responsible for the production of natural products, including pathway-specific transcriptional regulators like SARPs, are typically located in contiguous regions of the genome known as "biosynthetic gene clusters" (BGCs). The aim of this study was to activate expression of antibiotic BGCs in selected actinomycetes strains upon heterologous expression of the SARP-type regulator PapR2 from *Streptomyces pristinaespiralis*. Here, it was shown that PapR2 activates the undecylprodigiosin (Red) BGC in *Streptomyces coelicolor* A3(2) and thereby substitutes for RedD, the native SARP regulator of Red biosynthesis. In this study, overexpression of *papR2* increased the production of predominantly unknown antimicrobial compounds in more than half of the selected actinomycetes strains, as observed by bioassays against different test strains including bacteria and fungi. For strain *Streptomyces* sp. TÜ4106 it was found that *papR2* expression is associated with increased bioactivity and the production of a so far unknown blue pigmented substance. In summary, this study confirmed that activation of antibiotic BGCs can be successfully achieved by heterologous expression of SARP family regulators, thereby representing a powerful approach for novel bioactive natural product discovery.

Zusammenfassung

Aktinomyceten sind Gram-positive Bakterien, von denen mehrere Vertreter dafür bekannt sind eine Vielzahl von bioaktiven Naturstoffen zu produzieren, wie z. B. Antibiotika, Fungizide, antitumorale Agenzien oder Immunsuppressiva. Transkriptionsfaktoren aus der Familie der "*Streptomyces* antibiotic regulatory protein" (SARP) sind unter den Aktinomyceten, insbesondere bei den Streptomyceten, weit verbreitet und aktivieren bekanntermaßen die Antibiotika-Biosynthese. Die für die Antibiotika-Biosynthese verantwortlichen Gene, darunter auch die Gene, die für SARPs kodieren, liegen in der Regel gruppiert in spezifischen Regionen des Genoms vor, den sogenannten "Biosynthese-Genclustern" (BGCs). Ziel dieser Arbeit war es, die Antibiotika-Biosynthese in ausgewählten Aktinomyceten Stämmen durch heterologe Expression des SARP-Regulators PapR2 aus *Streptomyces pristinaespiralis* zu aktivieren. Es konnte gezeigt werden, dass PapR2 spezifisch das Undecylprodigiosin (Red)-BGC in *Streptomyces coelicolor* A3(2) aktiviert und damit RedD, den nativen SARP-Regulator der Red-Biosynthese, ersetzt. In der vorliegenden Arbeit wurde gezeigt, dass die Expression von *papR2* die Produktion von überwiegend unbekannten antimikrobiellen Substanzen in mehr als der Hälfte der ausgewählten Aktinomyceten Stämmen aktiviert, was durch Bioassays mit verschiedenen bakteriellen und pilzlichen Teststämmen festgestellt werden konnte. Für den Stamm *Streptomyces* sp. TÜ4106 wurde infolge der Expression von *papR2* eine erhöhte Bioaktivität verbunden mit der Produktion einer bisher unbekannten, blau pigmentierten Substanz nachgewiesen. Zusammenfassend wurde in dieser Studie gezeigt, dass Antibiotika-BGCs erfolgreich durch heterologe Expression von SARP-Transkriptionsfaktoren aktiviert werden können. Dies stellt einen leistungsstarken Ansatz für die Entdeckung neuer bioaktiver Naturstoffe dar.

Contents

Abbreviations

ACN	Acetonitrile
Act	Actinorhodin
Apra	Apramycin
APS	Ammonium persulfate
BGC	Biosynthetic Gene Cluster
BPB	Bromophenol blue
BTAD	Bacterial transcriptional activation domain
BTH	Bacterial Two Hybrid
CAM	Chloramphenicol
Carb	Carbenicillin
CIP	Ciprofloxacin
DBD	DNA binding domain
DMF	Dimethylformamide
DTT	Dithiothreitol
EDTA	Ethylenediaminetetraacetic acid
EMSA	Electrophoretic mobility shift assay
EtAc	Ethyl Acetate
Fos	Fosfomycin
GYM	Glucose Yeast Maltose
HEPES	4-(2-hydroxyethyl)-1-piperazineethanesulfonic acid
HPLC	High performance liquid chromatography
HTH	Helix – turn – helix
HZI	Helmholtz Centre for Infection Research
IBWF	"Institut für Biotechnologie und Wirkstoff-Forschung"
IMG	Integrated Microbial Genomes
JGI	Joint Genome Institute

Kan	Kanamycin
LAL	large ATP-binding regulators of the LuxR family
LB	Lennox Broth
MeOH	Methanol
NB	Nutrient Broth
NCBI	National Center for Biotechnology Information
NP	Natural product
NRPS	Nonribosomal peptide synthetase
ORF	Open reading frame
OSMAC	One strain many compounds
PAGE	Polyacrylamide gel electrophoresis
PEG	Polyethylene glycol
PKS	Polyketide synthase
Red	Undecylprodigiosin
SARP	*Streptomyces* antibiotic regulatory protein
SDS	Sodium dodecylsulfate
SFM	Soya Flour Mannitol
TAE	Tris Acetate Ethylenediaminetetraacetic acid
TBE	Tris/Borate/EDTA
TEMED	Thermo Scientific Pierce Tetramethylethylenediamine
WT	Wild type

List of Figures

List of Tables

Introduction

1

The discovery of bioactive molecules originating from microorganisms has been the greatest medical advancement in history. Over 70% of all antibiotics in clinical use are derived from natural products produced by the bacterial group of actinomycetes (Bérdy, 2012; Newman & Cragg, 2012). Paul Ehrlich introduced the first antibiotic in 1910, an arsenic-based synthetic drug called "salvarsan" to treat syphilis (Hutchings et al., 2019). Later Alexander Fleming discovered penicillin in 1928 as the first antibiotic of biological origin from the fungus *Penicillium notatum* (Hutchings et al., 2019). The first reports of antibiosis by actinomycetes appeared during the 1920s and the first antibiotic isolated from an actinomycete was actinomycin in 1940 by Waksman and Woodruff (Singh et al., 2010; Waksman & Woodruff, 1940). Since then, new antibiotics were found almost yearly during the Golden Age of antibiotic discovery from 1940s to 1960s (Hutchings et al., 2019). However, after the 1970s the discovery of new antibiotics has been declining gradually due to the frequent rediscovery of already known substances. Since the 'golden age' antibiotics have been extensively and uncontrollably applied to treat or prevent infections in not only humans but also pets, livestock, and crop, which led to increasing resistance of microorganisms to antibiotics (Bérdy, 2012; Hutchings et al., 2019). The Antibiotic Resistance Threats Report from 2019 estimated 18 antimicrobial-resistant bacteria and fungi in the US (CDC U.S., 2019). The most urgent threats originate from antibiotic-resistant bacteria such as carbapenem-resistant *Acinetobacter*, carbapenem-resistant Enterobacteriaceae and *Clostridioides difficile*, which is related to multiple antibiotic resistances. But also the commonly known pathogen methicillin-resistant *Staphylococcus aureus* (MRSA) constitutes a serious threat. It was estimated that each year antibiotic-resistant bacteria and fungi cause at least 2.8 Mio infections and almost 40,000 deaths per year, of which ~ 10% of infections and ~ 30% of deaths are attributed to *C. difficile* (CDC U.S., 2019). But also the risk for pandemics, i.e. large-scale outbreaks of infectious diseases, has been increased over the past century caused by an increase in global travel, urbanization, and of course, climate change (Madhav et al., 2017). Especially viral infections become more common because of the increasing probability for interspecies transmission (Carlson et al., 2022). The Covid-19 pandemic has shown how new viral diseases without established cures can cause significant social, economic, and political disruption. Furthermore, it has demonstrated that severe viral infections are often accompanied by a higher risk for the development of secondary bacterial infections (Bruyn et al., 2022). Thus, the discovery of new antibiotics and antivirals is urgently needed to combat the arising global threat of emerging infectious diseases and antibiotic resistances.

1.1   Antibiotics

Antibiotics were originally defined as low-molecular-weight substances (< 2 kDa) of biological origin that are biosynthesized in a step-wise manner and inhibit the growth of other microorganisms at low concentrations (< 200 µg/mL) (Waksman & Fenner, 1949). This definition was later modified to substances of diverse chemical structures that have inhibitory activity against microorganisms, viruses, and eukaryotic cells and also include synthetically or semi-synthetically produced antibiotics (Lancini & Lorenzetti, 1993). Some examples include polymyxin produced by *Bacillus* species, streptomycin and tetracyclines produced by *Streptomyces* species, and gentamicin produced by *Micromonospora purpurea* (Russell, 2004). Antibiotics can be classified into different categories based on their molecular structures, biosynthetic origin, mode of action, or target organisms (Schwalbe et al., 2007).

The most common classes for antibiotics divided by molecular structures include macrolides, tetracyclines, quinolones, aminoglycosides, peptide antibiotics, glycopeptides, and β-lactams, which are produced via different biosynthetic pathways. Antibiotics belonging to β-lactams and glycopeptides such as penicillin and vancomycin, respectively, are synthesized by non-ribosomal peptide synthetases (NRPS). Peptide antibiotics are either produced via the ribosomal peptide biosynthesis or by NRPSs. Erythromycin, which belongs to the class of macrolides, is synthesized by type-I polyketide synthases (PKS), whereas tetracyclines are synthesized by type-II polyketide synthases (Katz & Baltz, 2016). Antibiotics can also be classified according to their mode of action by which their selective toxicity is exerted upon the targets. The following five molecular sites have been identified as the major target sites of antibiotic action: cell wall, cell membrane, protein synthesis, nucleic acid synthesis, and key metabolic pathways (mostly folate metabolism) (Etebu & Arikekpar, 2016; Russell, 2004). Antibiotics targeting the cell wall of microorganisms such as β-lactams specifically inhibit the peptidoglycan biosynthesis resulting in cellular destabilization and ultimately in bursting of the bacterial cells due to osmotic pressure (Russell, 2004). The synthetic antibiotic isoniazid targets the synthesis of mycolic acid, which is an essential component of the cell membrane in mycobacteria and thereby is the treatment of choice for tuberculosis (Russell, 2004). Macrolides, tetracyclines, aminoglycosides, and chloramphenicol were found to inhibit protein biosynthesis in bacteria via different molecular mechanisms. For instance, erythromycin inhibits the 50 S ribosomal subunit by blocking the translocation during the elongation phase (Russell, 2004), while tetracyclines bind the 30 S ribosomal subunit and block the attachment of the tRNA (Riedel et al., 2019). All quinolones such as ciprofloxacin were reported to inhibit microbial DNA synthesis specifically by blocking DNA gyrases, which are essential for DNA replication and repair (Riedel et al., 2019). Lastly, sulphonamides are inhibitors of the folate metabolism as they block the synthesis of the precursor dihydrofolic acid by competitive inhibition of the dihydrofolic acid reductase (Russell, 2004).

Based on the mode of action of antibiotics, bacteria have developed different resistance mechanisms. Resistance mechanisms against β-lactams are especially well studied, which include synthesis of penicillin-destroying enzymes called β-lactamases that disrupt the β-lactam ring and thereby inactivate the antibiotic activity (Riedel et al., 2019). Other mechanisms of resistance that have been investigated are change of cellular permeability, alteration of the target, increased efflux, or metabolic bypass (Riedel et al., 2019). Cellular impermeability was especially observed in gram-negative bacteria, whereby the outer membrane provides a natural barrier and transporter molecules have been modified to resist antibiotic uptake (Delcour, 2009). In contrast to impermeability, some bacteria increasingly synthesize efflux pumps that are able to either excrete only one drug, a class of drugs or a wide range of substrates with only little or no chemical similarity (Russell, 2004). Alteration of the target has been observed in trimethoprim-resistant bacteria as they have developed a less resistant dihydrofolic acid reductase (Riedel et al., 2019). Metabolic bypass was also observed regarding the folate metabolism in sulphonamide-resistant bacteria, which acquired the ability to take up preformed folic acid and by that bypass the synthesis of the precursor dihydrofolic acid (Riedel et al., 2019). In summary, there is a large diversity among antibiotics regarding chemical structure, biosynthetic origin, and mode of action, and there still might be several antibiotic classes that were not discovered yet.

1.2   Actinomycetes and Secondary Metabolism

Actinomycetes are Gram-positive soil bacteria, of which most representatives have a mycelial lifestyle and a high G + C DNA content (Barka et al., 2016). Many actinomycetes, especially streptomycetes, are well-known for their ability to produce bioactive secondary metabolites and are responsible for the production of two-thirds of all commercially available antibiotics and other biotechnologically, industrially, or environmentally useful compounds, e.g. herbicides, pesticides, antitumor agents, or immunosuppressants (Barka et al., 2016; Tanaka & Omura, 1990). In general, the products of secondary metabolism are per definition not essential for survival or growth of the producing microorganisms (Russell, 2004). Secondary metabolism has been correlated with morphological differentiation, when under nutrient deprivation or minimal conditions, the vegetative mycelium differentiates into aerial hyphae. This moment often correlates with the peak of antibiotic production, when streptomycetes are grown on solid media (Barka et al., 2016). In liquid culture, antibiotic production coincides with the stationary phase of growth (Bibb, 2005).

The genes required for the regulation and synthesis of natural products as well as resistance are typically organized in contiguous regions referred to as "biosynthetic gene clusters" (BGCs). Examples include the BGCs from *Streptomyces coelicolor* A3(2), encoding the biosynthesis of the blue colored secondary metabolite actinorhodin (Act), the red colored undecylprdigiosin (Red), the lipopeptide calcium-dependent antibiotic (CDA), and the cryptic polyketide (CPK), which was later identified as "coelimycin" (Gomez-Escribano et al., 2012; Liu et al., 2013). Genomic studies of *S. coelicolor* A2(3) revealed that the genome harbors 23 BGCs for secondary metabolism in total (Bentley et al., 2002). A similar observation was reported for *S. avermitilis* as an even higher number of 30 BGCs related to secondary metabolism were identified, which is ~ 10-times higher than the number of previously known secondary metabolites (Ikeda et al., 2003). This led to the hypothesis that the antibiotic-producing capabilities of streptomycetes are higher than originally thought. Quantitative analysis of microbial genomes revealed that many actinomycetes with large genomes ranging from 3.64 to 12.7 Mb encode 20—50 BGCs and by that devote 0.8 to 3.0 Mb of coding capacity to secondary metabolite production (Baltz, 2017). A large-scale bioinformatic analysis of ~ 170,000 bacterial genomes and ~ 47,000 metagenome assembled genomes disclosed that "only 3% of the natural products potentially encoded in bacterial genomes have been experimentally characterized" (Gavriilidou et al., 2022). The difference in the amount of potentially produced natural products and actually identified natural products was attributed to the vast majority of natural products not being expressed under standard laboratory conditions (Baltz, 2017). The respective BGCs were referred to as "silent", although gene expression can occur at a very low level. It was suggested that~90% of BGCs contained within a microbial genome are silent (Baltz, 2017). Activation of these silent BGCs has become a key objective in the search for novel natural products and can be achieved by applying state-of-the-art technology.

1.3   Activation of Antibiotic Biosynthetic Gene Clusters

Different technologies and experimental approaches can be applied to either activate BGC expression in an untargeted manner or to target a specific BGC. Untargeted activation is based on the idea to provide different conditions in order to activate secondary metabolism. For instance, the "one strain many compounds" (OSMAC) approach involves variation of nutrient levels and other parameters and is based on the knowledge that even small changes of cultivation parameters can have a large effect on secondary metabolism (Covington et al., 2021). For instance, a great increase in penicillin production was observed when xylose was used as carbon source instead of glucose (Soltero. & Johnson., 1953). Furthermore, since natural products are assumed to have protective, predative or communicative purposes, the addition of other microorganisms may also lead to the activation of secondary metabolism. For instance, co-cultivation of *S. coelicolor* A3(2) with the predatory bacterium *Myxococcus xanthus* led to the production of actinorhodin in *S. coelicolor* A3(2) as well as the siderophores desferrioxamine B and myxochelin, respectively, while both bacteria were competing for iron (Lee et al., 2020). Furthermore, it was discovered that the addition of small molecules and even clinical antibiotics activate secondary metabolism as well (Okada & Seyedsayamdost, 2017). Based on this knowledge the high-throughput elicitor screen (HiTES) technology has been developed, where a library of small molecules is applied. This approach revealed that the clinical antibiotic trimethoprim is a global activator of secondary metabolism in *Burkholderia thailandensis*, where it induced the production of over 100 compounds and even led to identification of a new family of metabolites called acybolins (Okada et al., 2016). Untargeted approaches have the advantage of being genetics-free and can be applied to even less-studied microbes that may not be genetically tractable. However, these experiments are often accompanied with large screening processes and time consuming analyses (Covington et al., 2021).

Targeted approaches are applied to activate expression of a specific BGC, which was often identified with bioinformatic analyses, typically followed by genetic manipulation of the microbe. For instance, the publicly available tool anti-SMASH identifies BGCs within microbial genomes by searching for co-localized genes encoding biosynthetic enzymes such as PKSs or NRPSs (Blin et al., 2021). The expression of biosynthetic gene clusters often relies on pleiotropic or pathway-specific transcriptional regulators, which respond to different, mostly unknown stimuli (Wohlleben et al., 2017). That a BGC is silent is frequently due to inactivity of positive transcriptional regulators or repression by negative transcriptional regulators. Expression of a BGC can be enforced by placing it under control of constitutively active or inducible promoters. An artificial promoter, which is frequently used for homologous and heterologous expression in actinomycetes is a derivative of the promoter of the erythromycin resistance gene *ermE* from *Saccharopolyspora erythraea* (*ermE*p*), which is a constitutively active promoter (Bibb et al., 1994). Another well-known promoter is the thiostrepton-inducible promoter *tipA*p, which is a commonly used inducible promoter that allows to control the expression of the BGC of interest at a certain time point upon addition of thiostrepton as an inducer (Liu et al., 2021). Genetic manipulation of the regulatory machinery is another possible approach for activation of BGC expression by overexpression of transcriptional activators or knocking out transcriptional repressors (Wohlleben et al., 2017). Overexpression of transcriptional activators can be achieved by placing the gene expression of the activator under control of artificial promoters. In *Streptomyces ambofaciens* the constitutive expression of a large ATP-binding regulator of the LuxR family (LAL) protein under control of *ermE*p* induced expression of a silent PKS BGC and led to identification of stambomycins A-D (Laureti et al., 2011). Deletion of *gbnR*, which codes for a putative transcriptional repressor in *S. venezuelae*, resulted in derepression of a silent BGC and thereby discovery of gaburedins (Sidda et al., 2014). However, introducing an artificial promoter in front of a BGC, or manipulation of cluster-situated transcriptional regulators are completely restricted to only one BGC (Wohlleben et al., 2017). Simultaneously, application of untargeted approaches such as OSMAC and HiTES are completely unspecific in terms of activation of BGC expression (Wohlleben et al., 2017). An approach that compromises between these two issues is the expression of non-native transcriptional activators. For example, heterologous expression of the gene encoding pathway-specific PAS-LuxR regulator PimM from *S. natalensis* led to activation of clavulanic acid, cephamycin C, and tunicamycin BGCs in *S. clavuligerus* (Martínez-Burgo et al., 2019). Thus, heterologous expression of conserved, pathway-specific transcriptional regulators in foreign actinomycetes species is a promising activation strategy as it targets a defined set of BGCs (Krause et al., 2020).

1.4   SARP Regulators

Streptomyces antibiotic regulatory protein (SARP) family regulators are pathway-specific activators of antibiotic biosynthesis found in actinobacteria, mostly streptomycetes (Wohlleben et al., 2017). Examples of SARP-type regulators are RedD and ActII-ORF4, which are activators of Red and Act biosynthesis in *S. coelicolor* A3(2), respectively, DnrI activating daunorubicin biosynthesis in *Streptomyces peucetius* (Cundliffe, 2006), TylS and TylT, which both activate tylosin production in *Streptomyces fradiae* (Bate et al., 1999), or PapR1, PapR2, and PapR4 activating pristinamycin synthesis in *Streptomyces pristinaespiralis* (Mast et al., 2015). Besides the ability to activate expression of BGCs, very little is known about SARPs regarding the protein structure, possible triggers, and molecular mechanisms of transcriptional activation. However, crystallization of proteins similar to SARPs such as EmbR of *Mycobacterium tuberculosis* enabled predictions for the protein structure of SARPs (Alderwick et al., 2006). Based on these studies it was suggested that SARPs contain an N-terminal winged helix-turn-helix (HTH) DNA binding domain (DBD) that mostly binds heptameric tandem repeats ~ 8 bp upstream of the—10 promoter element, which overlap the—35 promoter region (Liu et al., 2013; Wietzorrek & Bibb, 1997). In the case of actinorhodin and daunorubicin regulation it was observed that the tandem repeats with the consensus sequences: 5'-TCGAGCG/C and 5'-TCGAGCG bound by SARPs are separated by 11 or 22 bp (including the 7 bp motif itself) corresponding to one or two turns of the DNA helix (Arias et al., 1999; Sheldon et al., 2002; Wietzorrek & Bibb, 1997). Mechanistic studies on the SARP regulator AfsR suggested a model for SARP transcriptional activation as two monomers binding to the same face of the DNA while the RNA polymerase binds to the opposite site of the helix resulting in a DNA-(SARP)$_2$-RNAP transcription initiation complex (Tanaka et al., 2007).

SARPs can be divided according to size into three classes: "small" SARPs such as RedD, ActII-ORF4, and PapR2 are less than 300 residues long while "medium" and "large" SARPs, such as AfsR contain over 600 residues (Liu et al., 2013). All these three classes share three major functional domains: the N-terminal DBD, a central ATPase domain as bacterial transcriptional activation domain (BTAD), and a conserved C-terminal domain of unknown function. "Medium" and "large" SARPs mostly contain additional domains at the C-terminus (Liu et al., 2013). In a study of Li et al. (2018), it was shown that the SARP regulator NosP regulates the production of nosiheptide and senses its biosynthetic process by interacting with peptidyl and small ligands derived from a nosiheptide precursor peptide. This study provides first evidence of a SARP to respond to small-molecule ligands.

Due to many SARPs containing similar protein structures and binding to similar consensus sequences, they have been proposed as suitable tools to activate expression of silent BGCs in foreign actinomycetes species upon heterologous expression. The small SARP PapR2, which is essential for the production of pristinamycin in *S. pristinaespiralis* (Mast et al., 2015), was reported to induce expression of the silent Red BGC in *S. lividans* (Krause et al., 2020). This indicates that PapR2 took over the regulatory role of the native SARP RedD in *S. lividans* (Krause et al., 2020). The heterologous expression of a panel of SARP-type transcriptional activators including PapR2 in *Streptomyces* sp. CA-256286 led to the identification of a silent BGC encoding the polyketide griseusin A (Beck et al., 2021). Furthermore, heterologous expression of PapR2 in a novel Indonesian strain isolate *Streptomyces* sp. SHP22-7 led to the activation of an amicetin/plicacetin gene cluster (Krause et al., 2020). These experiments verify the potential of SARPs, especially PapR2, to activate BGCs by heterologous expression.

1.5   Aim of the Study

This study aims to elucidate the potential of the SARP-type regulator PapR2 from *Streptomyces pristinaespiralis* to be used as a general activator of BGC expression in different actinomycetes. For this purpose, genome-sequenced actinomycetes from the DSMZ strain collection will be prioritized, through bioinformatics analysis of the genome, as candidate strains for *papR2* expression. *papR2* overexpression constructs, suitable for broad application in different actinomycete strains, will be generated and tested for their functionality in the model organism *Streptomyces coelicolor*. Subsequently, the *papR2* overexpression constructs will be transferred to the selected DSMZ actinomycetes strains. The effect of PapR2-specific BGC activation on secondary metabolite production will be assessed via bioactivity assays against a panel of test organisms. Furthermore, PapR2 derivatives obtained from expression in *E. coli* of codon-optimized *papR2* variants will be assessed for their DNA binding capacity as preparative studies for subsequent protein crystallization and mechanistic studies of PapR2.

Material and Methods

2

2.1 Material

2.1.1 Microorganisms

Table 2.1 Bacterial strains used in this study

Strain	Genotype	Origin/Reference
Escherichia coli		
DH5α (NEB® 5-alpha)	*fhuA2 Δ(argF-lacZ)U169, phoA, glnV44, Φ80, Δ(lacZ)M15, gyrA96, recA1, relA1, endA1, thi-1, hsdR17*	New England Biolabs
ET12567	*F-, dam-13::Tn9, dcm-6, hsdM, hsdR, zjj-202::Tn10, recF143, galK2, galT22, ara-14, lacY1, xyl-5, leuB6, thi-1, tonA31, rpsL136, hisG4, tsx-78, mtl-1, gln*	MacNeil et al. (1992)

(continued)

Supplementary Information The online version contains supplementary material available at https://doi.org/10.1007/978-3-658-44552-2_2.

11

Table 2.1 (continued)

Strain	Genotype	Origin/Reference
ET12567 (puZ8002)	ET12567 carrying pUZ8002, an RK2 derivative obtained by Wilson, J. and Figurski, D. H.; pUZ8002 supply transfer functions.	Paget et al. (1999)
Streptomyces		
S. coelicolor A3(2) M145	A3(2) (SCP1⁻, SCP2⁻)	Hopwood et al. (1995)
S. coelicolor M510	M145 $\Delta redD$	Floriano & Bibb (1996)
S. coelicolor M512	M145 $\Delta redD$, ΔactII-ORF4	Floriano & Bibb (1996)
S. coelicolor M513	M145 $\Delta afsR$	Floriano & Bibb (1996)
S. coelicolor M510 (pGM190)	M145 $\Delta redD$, *tipAp, tsr, aphII*	This study
S. coelicolor M512 (pGM190)	M145 $\Delta redD$, ΔactII-ORF4, *tipAp, tsr, aphII*	This study
S. coelicolor M513 (pGM190)	M145 $\Delta afsR$, *tipAp, tsr, aphII*	This study
S. coelicolor M510 (pGM190/*papR2*)	M145 $\Delta redD$, *tipAp, tsr, aphII, papR2*	This study
S. coelicolor M512 (pGM190/*papR2*)	M145 $\Delta redD$, ΔactII-ORF4, *tipAp, tsr, aphII, papR2*	This study
S. coelicolor M513 (pGM190/*papR2*)	M145 $\Delta afsR$, *tipAp, tsr, aphII, papR2*	This study
S. coelicolor M510 (pRM4)	M145 $\Delta redD$, , ΔactII-ORF4, *ermEp*, aac(3)IV*, artificial *RBS*	This study
S. coelicolor M512 (pRM4)	M145 $\Delta redD$, *ermEp*, aac(3)IV*, artificial *RBS*	This study
S. coelicolor M513 (pRM4)	M145 $\Delta afsR$, *ermEp*, aac(3)IV*, artificial *RBS*	This study

(continued)

Table 2.1 (continued)

Strain	Genotype	Origin/Reference
S. coelicolor M510 (pRM4/*papR2*)	M145 Δ*redD, erm*Ep*, *aac(3)IV*, artificial *RBS, papR2*	This study
S. coelicolor M512 (pRM4/*papR2*)	M145 Δ*redD, erm*Ep*, *aac(3)IV*, artificial *RBS, papR2*	This study
S. coelicolor M513 (pRM4/*papR2*)	M145 Δ*afsR, erm*Ep*, *aac(3)IV*, artificial *RBS, papR2*	This study
S. coelicolor M510 (pGM1190)	M145 Δ*redD, tip*Ap, *tsr, aac(3)IV*	This study
S. coelicolor M512 (pGM1190)	M145 Δ*redD, ΔactII-ORF4, tip*Ap, *tsr, aac(3)IV*	This study
S. coelicolor M513 (pGM1190)	M145 Δ*afsR, tip*Ap, *tsr, aac(3)IV*	This study
S. coelicolor M510 (pGM1190/*papR2-tip*Ap)	M145 Δ*redD, tip*Ap, *tsr, aac(3)IV, papR2*	This study
S. coelicolor M512 (pGM1190/*papR2-tip*Ap)	M145 Δ*redD, ΔactII-ORF4, tip*Ap, *tsr, aac(3)IV, papR2*	This study
S. coelicolor M513 (pGM1190/*papR2-tip*Ap)	M145 Δ*afsR, tip*Ap, *tsr, aac(3)IV, papR2*	This study
S. coelicolor M510 (pGM1190/ *papR2-erm*Ep*)	M145 Δ*redD, erm*Ep*, *tsr, aac(3)IV, papR2*	This study
S. coelicolor M512 (pGM1190/*papR2-erm*Ep*)	M145 Δ*redD, ΔactII-ORF4, erm*Ep*, *tsr, aac(3)IV, papR2*	This study
S. coelicolor M513 (pGM1190/*papR2-erm*Ep*)	M145 Δ*afsR, erm*Ep*, *tsr, aac(3)IV, papR2*	This study
S. ambofaciens DSM 40697 (pGM190)	*tip*Ap, *tsr, aphII*	This study
S. ambofaciens DSM 40697 (pGM190/*papR2*)	*tip*Ap, *tsr, aphII, papR2*	This study
S. avermitilis DSM 46492	Wild type genotype	DSMZ strain collection

(continued)

Table 2.1 (continued)

Strain	Genotype	Origin/Reference
S. avermitilis DSM 46492 (pRM4/*papR2*)	*ermE*p*, *aac(3)IV*, artificial RBS, *papR2*	This study
S. avermitilis DSM 46492 (pGM1190/*papR2-tip*Ap)	*tip*Ap, *tsr*, *aac(3)IV*, *papR2*	This study
S. leeuwenhoekii DSM 42122	Wild type genotype	DSMZ strain collection
S. leeuwenhoekii DSM 42122 (pGM190/*papR2*)	*tip*Ap, *tsr*, *aphII*, *papR2*	This study
S. leeuwenhoekii M1653 (pGM1190)	Δ*cxmK*::*ne* (chaxamycin non-producing mutant), *tip*Ap, *tsr*, *aac(3)IV*	This study M1653 strain from Castro et al. (2015)
S. leeuwenhoekii M1653 (pGM1190/*papR2-tip*Ap)	Δ*cxmK*::*ne* (chaxamycin non-producing mutant), *tip*Ap, *tsr*, *aac(3)IV*, *papR2*	This study M1653 strain from Castro et al. (2015)
S. misionensis DSM 40306	Wild type genotype	DSMZ strain collection
S. misionensis DSM 40306 (pGM1190/*papR2-tip*Ap)	*tip*Ap, *tsr*, *aac(3)IV*, *papR2*	This study
S. platensis DSM 40041 (pGM190)	*tip*Ap, *tsr*, *aphII*	This study
S. platensis DSM 40041 (pGM190/*papR2*)	*tip*Ap, *tsr*, *aphII*, *papR2*	This study
Streptomyces sp. DSM 40976	Wild type genotype	DSMZ strain collection
Streptomyces sp. DSM 40976 (pGM1190/*papR2-tip*Ap)	*tip*Ap, *tsr*, *aac(3)IV*, *papR2*	This study
Streptomyces sp. TÜ4106 (pGM190)	*tip*Ap, *tsr*, *aphII*	Krause dissertation (2021)
Streptomyces sp. TÜ4106 (pGM190/*papR2*)	*tip*Ap, *tsr*, *aphII*, *papR2*	Krause dissertation (2021)
Streptomyces sp. TÜ4128 (pGM190)	*tip*Ap, *tsr*, *aphII*	Krause dissertation (2021)
Streptomyces sp. TÜ4128 (pGM190/*papR2*)	*tip*Ap, *tsr*, *aphII*, *papR2*	Krause dissertation (2021)

(continued)

Table 2.1 (continued)

Strain	Genotype	Origin/Reference
Streptomyces sp. TÜ4134 (pGM190)	*tip*Ap, *tsr, aphII*	Krause dissertation (2021)
Streptomyces sp. TÜ4134 (pGM190/*papR2*)	*tip*Ap, *tsr, aphII, papR2*	Krause dissertation (2021)
Kitasatospora		
K. cineracea DSM 44780	Wild type genotype	DSMZ strain collection
K. cineracea DSM 44780 (pRM4/*papR2*)	*erm*Ep*, *aac(3)IV*, artificial RBS, *papR2*	This study
K. cineracea DSM 44780 (pGM1190/*papR2-tip*Ap)	*tip*Ap, *tsr, aac(3)IV, papR2*	This study
K. niigatensis DSM 44781	Wild type genotype	DSMZ strain collection
K. niigatensis DSM 44781 (pRM4/*papR2*)	*erm*Ep*, *aac(3)IV*, artificial RBS, *papR2*	This study
K. niigatensis DSM 44781 (pGM1190/*papR2-tip*Ap)	*tip*Ap, *tsr, aac(3)IV, papR2*	This study

Table 2.2 Microorganisms used for bioactivity assays

Strain	Genotype	Origin/Reference
E. coli K12 (DSM 5698)	*F-, lambda-, ilvG-, rfb-50, rph-1*	DSMZ strain collection
E. coli A593 (*tolB2*)	*F-, thr-1, leuB6(Am), fhuA21, lacY1, glnX44(AS), tolB2, λ-, rfbC1, thiE1*	*E. coli* genetic stock center, Yale https://cgsc.biology.yale.edu/Strain.php?ID=7869
E. coli PB3 (*tolC5(del)*)	*F-, lacY1 or lacZ4, gal-6, hisG1, ΔtolC5, uxaC201, rpsL8 or rpsL104 or rpsL17, malT1(λ^R)?, mtlA2?*	*E. coli* genetic stock center, Yale https://cgsc.biology.yale.edu/Strain.php?ID=10140
E. coli D22 (*lpxC101*)	*F-, lpxC101, proA23, lac-28, tsx-81, trp-30, his-51, tufA1, rpsL173(strR), ampCp-1*	*E. coli* genetic stock center, Yale https://cgsc.biology.yale.edu/Strain.php?ID=10019
Botrytis cinerea	Wild type genotype	DSMZ strain collection
Micrococcus luteus	Wild type genotype	DSMZ strain collection
Saccharomyces cerevisiae	Wild type genotype	DSMZ strain collection

2.1.2   Vectors and Plasmids

Table 2.3   Vectors and plasmid constructs used in this study

Plasmid/vector	Properties	Reference
pET28(+)_papR2	pET28(+) carrying codon-optimized complete papR2 sequence, *lacI, pBR322 ori, aphII*	K. Büssow, HZI
pET28(+)_papR2_257	pET28(+) carrying codon-optimized truncated papR2 sequence, *lacI, pBR322 ori, aphII, papR2-257* derivative	K. Büssow, HZI
pGM190	Shuttlevector, *tipAp, tsr, aphII, rep*-ts	Muth (2018)
pGM190/*papR2*	*tipAp, tsr, aphII, rep*-ts, *papR2*	Mast et al. (2015)
pRM4/*papR2*	pSET152, *ermE*p* derivative (Φ C31 integration vector) *aac(3)IV*, artificial *RBS, papR2*	Krause et al. (2020)
pGM1190	*tipAp, tsr, aac(3)IV, rep, sso, dso*	Muth (2018)
pGM1190/*papR2-tipAp*	*tipAp, tsr, aac(3)IV, rep, sso, papR2*	This study
pGM1190/*papR2-ermE*p*	*ermE*p*, tsr, aac(3)IV, rep, sso, papR2*	This study

2.1.3   Oligonucleotides

Table 2.4   Oligonucleotides used for sequencing of generated plasmids (by Eurofins Genomics)

Primer	Sequence	Plasmid	Origin
M13R - 22mer	TCA CAC AGG AAA CAG CTA TGA C	pGM1190; pGM1190/ *papR2-tipAp*	Gomez-Escribano et al. (2019)
pIJ10257_Nde_1	CGA GTG TCC GTT CGA GTG	pGM1190/ *papR2-ermE*p*	J.P. Gomez-Escribano (pers. communication)

2.1.4   Media

Table 2.5   Media used in this study. 16 g/L agar used for solid media if not stated otherwise

Medium	Ingredients (per liter of final volume)
GYM	4 g Glucose 4 g Yeast extract 10 g Malt extract For agar medium: 12 g/L agar 2 g $CaCO_3$ pH 7.2
HM	10 g Malt extract 5 g Yeast extract 3 g Glucose pH 7.3
KIV	20 g Malt extract pH 6.0
LB (Lennox, 1955)	10 g Tryptone, 5 g Yeast extract, 4 g NaCl, (1 g Glucose), pH 7.3
NB soft agar	8 g Nutrient Broth, 5 g Agar
Oat meal (OM)	20 g Oat meal, gluten-free (Seitz) Trace element 5 mL pH 7.3

(continued)

Table 2.5 (continued)

Medium	Ingredients (per liter of final volume)
R5	103 g Sucrose 10 g Glucose 0.25 g K_2SO_4 10.12 g $MgCl_2$ 0.1 g Casamino acids 5 g Yeast extract 5.73 g TES 2 ml Trace element solution 18 g Agar pH 7.2 Added after autoclaving: 20 ml 1 M $CaCl_2$ 10 ml 0.54% KH_2PO_4 15 ml 20% L-Prolin
S (modified from Okanishi et al., 1974)	4 g Peptone 4 g Yeast extract 4 g K_2HPO_4 2 g KH_2PO_4 10 g Glycine in 800 mL H_2O Added after autoclaving: 10 g Glucose 0.5 g $MgSO_4$ in 200 mL H_2O
SFM	20 g Soy flour (full fat) 20 g Mannitol
YT (2X)	16 g Tryptone 10 g Yeast extract 5 g NaCl pH 7.0

2.1.5 Buffers

Table 2.6 Buffers used in this study. All volumes and quantities apply to 1 L

Buffer	Ingredients
CaCl$_2$-solution	50 mM CaCl$_2$
P buffer (Thompson et al., 1982)	10.3% Sucrose 2% Trace element 25 mM TES (pH 7.2) 1.4 mM K$_2$SO$_4$ 10 mM MgCl$_2$ 0.4 mM KH$_2$PO$_4$ 2.5 mM CaCl$_2$
Transformation buffer	P buffer with 25% PEG
Trace element solution	200 mg Fe x 6 H$_2$O 10 mg Na$_2$B$_4$O$_7$ x 10 H$_2$O 10 mg (NH$_4$)$_6$Mo$_7$O$_{24}$ x 4 H$_2$O 10 mg CuCl$_2$ x 2 H$_2$O 10 mg MnCl$_2$ x 4 H$_2$O 40 mg ZnCl$_3$
TAE buffer	40 mM Tris Base 20 mM acetic acid 1 mM EDTA pH 8.6

2.1.6 Antibiotics

Table 2.7 Antibiotic stock solutions and supplements to selection media

Antibiotic	Concentration
Apramycin (Apra)	50 μg/mL
Carbenicillin (Carb)	100 μg/mL
Chloramphenicol (CAM)	25 μg/mL in 100% Ethanol
Fosfomycin (Fos)	25 μg/mL
Kanamycin (Kan)	50 μg/mL

2.1.7 Solutions for SDS Polyacrylamide Gel Electrophoresis (SDS-PAGE) and Bandshift Assay

Table 2.8 Solutions and buffers used for SDS PAGE and agarose bandshift assays

Solutions	Ingredients
Binding Buffer (5X) (Bandshift assay)	100 mM HEPES (pH 7.6) 5 mM EDTA 50 mM $(NH_4)_2SO_4$ 5 mM DTT 1% Tween20 150 mM KCl
Stacking gel buffer	0.5 M Tris/HCl 0.4% SDS pH 6.8
Stacking gel (6%)	3 mL Acrylamide 3.75 mL stacking gel buffer 45 µL APS (10%) 13.5 µL TEMED 8.25 mL H_2O
Decolorizing solution	20% Ethanol 10% Acetic acid (100% stock)
Loading buffer (without Bromophenol blue) (Bandshift assay)	0.25X TBE Buffer 60% Glycerol
Protein sample buffer (2X)	0.125 M Tris/HCl (pH 6.8) 4% SDS 20% Glycerol 2 mM EDTA 0.02% Bromphenol blue 3% DTT
Resolving gel buffer	1.5 M Tris/HCl 0.4% SDS pH 8.8
Resolving gel	12.5 mL Acrylamide 7.5 mL Resolving gel buffer 100 µL APS (10%) 25 µL TEMED 10 mL H_2O
1X TBE Buffer (Bandshift assay)	12.1 g Tris 5.13 g Borate 0.37 g EDTA
Tris glycine buffer (10X)	0.25 M Tris 1.92 M Glycine 1% SDS

2.1.8 Enzymes, Markers, Kits, and Other Material

2.1.8.1 Enzymes

Table 2.9 Enzymes used in this study

Enzyme	Manufacturer
AseI	New England Biolabs
NdeI	New England Biolabs
EcoRI	New England Biolabs
Lysozyme (from chicken egg white)	Sigma Aldrich
T4 ligase	New England Biolabs

2.1.8.2 DNA and Protein Marker

Table 2.10 DNA and protein markers used in this study

Marker	Size	Manufacturer
1 kB DNA Ladder	10002, 8001, 6001, 5001, 4001, 3001, 2000, 1500, 1000, 517, 500 bp	New England Biolabs
Prestained Protein Ladder (10–180 kDA)	180, 135, 100, 75, 63, 48, 35, 25, 17, 11 kDa	BioFroxx

2.1.8.3 Kits

Table 2.11 Kits used in this study for DNA preparation

Kit	Purpose	Manufacturer
Wizard Plus SV Minipreps DNA Purification System	DNA extraction from bacterial cells	Promega
Wizard SV Gel and PCR Clean-Up System	DNA purification from gel or PCR	Promega

2.1.8.4 Other Material

Ciprofloxacin discs (5 μg)	Microexpress
Coomassie Protein Assay Reagent	Thermo Scientific
CutSmart Buffer (10X)	New England Biolabs
HDGreen Plus	Intas
Purple loading dye (6X)	New England Biolabs
ROTIPHORESE®NF-Acrylamid/Bis-solution	Roth
T4 ligase Buffer (10X)	New England Biolabs

2.2   Methods

2.2.1   Bioinformatic Analyses

To find DSMZ strains whose genomes have been sequenced and carry candidate BGCs for activation by PapR2, an advanced bioinformatic search for candidate genomes was conducted with all available genomes from the Integrated Microbial Genomes (IMG) facility of the Joint Genome Institute (JGI) (Chen et al., 2019). BGCs that contain genes coding for SARP transcriptional regulators were identified with antiSMASH 6.0 (Blin et al., 2021) and "relaxed" was set as strictness. The genomes were drawn from National Center for Biotechnology Information (NCBI, https://www.ncbi.nlm.nih.gov/). NCBI BLASTP was used for comparison of amino acid sequences of SARPs with the PapR2 amino acid sequence (see supplementary material). Structure predictions of PapR2 and other proteins were performed with AlphaFold and the Mol* 3D Viewer of RCSB Protein Data Bank (https://www.rcsb.org/) was used for visualization. I-TASSER (Iterative Threading ASSEmbly Refinement) was used for prediction of putative ligands of PapR2.

2.2.2   Cultivation and Storage of Microorganisms

2.2.2.1   Cultivation and Preservation of *E. coli*

E.coli strains (Tables 2.1, 2.2) were inoculated in 10–20 ml LB medium (Table 2.5) in Falcon tubes or flasks containing a 1/1000 dilution of appropriate antibiotics when required (Table 2.7). Cultures were incubated overnight at 37 °C under shaking at 220 rpm. Cultures were inoculated with a single, picked colony from solid LB medium, or with an aliquot of a glycerol stock (typically 10–50 μL). The growth was monitored by measuring the optical density using a spectrophotometer at OD_{600} nm (BioChrom). For cultivation on plate, 50–300 μL of *E. coli* cell suspension was evenly distributed on LB agar containing appropriate antibiotics when required and incubated at 37 °C overnight. *E. coli* strains were stored as glycerol stock containing 20% glycerol at −20 °C or −80 °C for longer term.

2.2.2.2   Cultivation and Preservation of *Streptomyces*

Agar blocks of a mature solid culture (size ~ 1 cm^2) or 10 μL of spores from glycerol stock of *Streptomyces* strains (Table 2.1) were used to inoculate 50 mL of liquid cultures with S-, GYM, or R5 medium (Table 2.5). Cultures were cultivated for 2–5 days in 250 mL flasks with baffles to increase aeration and dispersion under shaking at 150–180 rpm at 30 °C.

For cultivation on agar plates, mycelium from a grown plate or 10–50 μL spores from a glycerol stock were evenly distributed with a cotton tip on GYM, SFM, or R5 containing appropriate antibiotics and kept at 28 °C (Tables 2.5, 2.7). To obtain *Streptomyces* spores for storage, GYM and SFM plates containing a *Streptomyces* strain were incubated for 5–15 days at 28 °C. The spores were collected from the plates with a cotton tip, transferred into 1.5 mL of 20% glycerol and stored at −20 °C.

2.2.3   *In vitro* Manipulation and Analysis of DNA

2.2.3.1   Isolation of Plasmid DNA from *E. coli*

The alkaline lysis is a fast and prominent method for the extraction of plasmid DNA. This method was used to obtain higher amounts of plasmid DNA for cloning, transformation and conjugation purposes. For this purpose, an *E. coli* colony or 50–200 μL from a glycerol stock was used to inoculate 10 mL of LB medium containing appropriate antibiotics when required and incubated at 30 °C and 220 rpm overnight. Plasmid isolation was performed with the DNA Purification System Kit following the manufacturer's instructions (Table 2.11). In short, this included the following procedure: culture was centrifuged at 13300 rpm for 10 min and the pellet was resuspended in 250 μL P1 buffer. Then, 10 μL alkaline phosphatase and 250 μL P2 buffer were added. After repeated inversion, 350 μL P3 buffer was added and centrifuged at 13300 rpm for 1 min. The supernatant was transferred onto a column and washed twice with washing buffer containing ethanol under centrifugation at 13300 rpm for 1 min. After washing, the column was dried by repeated centrifugation at 13300 rpm for 1 min. DNA was eluted with 100–150 μL nuclease-free water. DNA concentration was quantified with a NanoDrop™ UV-Vis-spectrophotometer One/One c.

2.2.3.2 Digestion of DNA with Restriction Endonucleases
Plasmid DNA was digested with specific restriction endonucleases to either perform cloning experiments or confirm correct plasmid construction (Table 2.9). The restriction was performed under the conditions recommended by the suppliers of the endonucleases. The reaction mix contained the following components: 0.5–1 μL restriction enzyme, 100–1000 ng DNA, the required amount of supplied buffer (typically 10% v/v of CutSmart buffer), and filled up with nuclease-free water. The reaction volumes varied between 10–200 μL. The reaction mixture was incubated at the recommended temperature (typically 37 °C) for 1–4 h to ensure complete digestion and analyzed by gel electrophoresis.

2.2.3.3 Analysis of DNA Fragments by Agarose Gel Electrophoresis
An agarose gel electrophoresis serves as a method to separate DNA fragments according to size and topology. The generation of an electric field leads to the movement of the negatively charged DNA towards the positive pole, while the agarose allows the small molecules to travel faster. The restricted DNA products were resolved in an 0.8% agarose gel in 1X TAE buffer and stained with 10X HDGreen Plus (Table 2.6). The DNA samples were mixed with 6X purple loading dye and 500 ng of 1 kB DNA ladder was used as marker (Table 2.10). Gel electrophoresis was conducted for 35–100 min at a constant voltage of 95–110 V. DNA was visualized using a Blue-LED transilluminator in a Gel Stick "Touch" documentation system (Intas).

2.2.3.4 Purification of DNA Fragments from Agarose Gel
When gel electrophoresis was conducted to obtain specific DNA fragments for cloning procedures, the respective agarose blocks were excised using a scalpel and purified with the Wizard PCR and Gel purification Kit following manufacturer's instructions (Table 2.11). In short, the gel was dissolved with 500 μL membrane binding buffer at 60 °C and applied onto a column. The column was washed twice with washing buffer containing ethanol by centrifugation at 13300 rpm for 1 min. DNA was eluted with 20–40 μL nuclease-free water by centrifugation at 13300 rpm for 2 min.

2.2.3.5 Ligation of DNA Fragment with Cloning Vector

For ligation the purified vector and insert were applied in a 1:3 ratio to a 10–15 μL reaction mix containing 1–1.5 μL T4 ligase buffer, 1 μL T4 ligase and filled up with nuclease-free water (Table 2.9). To verify the sequence of synthesized DNA, DNA products were sent with appropriate oligonucleotides to Eurofins Genomics for Sanger Sequencing (Table 2.4).

2.2.4 DNA Transfer to Microorganisms

2.2.4.1 Preparation of Competent Cells of *E. coli*

The $CaCl_2$ method induces chemically bacterial competence enabling the uptake of free DNA (Cohen et al., 1972). This method includes the preparation of competent cells of *E. coli*. First, 50 mL of LB medium was inoculated with the required *E. coli* strain and cultivated overnight at 37 °C and 220 rpm. 1 mL of the overnight culture was used to inoculate fresh 50 mL of LB medium and this culture was incubated under the same conditions up to an OD_{600} of 0.5–0.8. The cells were harvested by centrifugation for 5 min at 4 °C and 5000 rpm. The pellet was resuspended in 10 mL of ice-cold sterile ultrapure H_2O and centrifuged for 5 min at 4 °C and 5000 rpm. The pellet was resuspended in 2 mL of 50 mM $CaCl_2$ + 15% glycerol (Table 2.6). Aliquots of 200 μL were prepared and stored at −80 °C or directly used for transformation.

2.2.4.2 Transformation of *E. coli* with $CaCl_2$ Method

To transform *E. coli* (Table 2.1) 5–10 μL (~ 500 ng) of plasmid DNA or ligation product (Table 2.3) were added to 200 μL competent cells and incubated on ice for 30 min. Then, cells were heat shocked for 45 sec at 42 °C to increase cell wall and membrane permeability and allow the entry of DNA. After heat shock, the tube was incubated on ice for 1 min. 1 mL of LB medium was added and cells were incubated for 60 min at 37 °C to initiate cell regeneration and expression of resistance genes. Afterwards, the transformation was used to inoculate plates of LB agar containing appropriate antibiotics. The plates were incubated overnight at 37 °C.

2.2.4.3 Preparation of Protoplasts of Actinomycetes

Protoplast transformation is a commonly used and early established procedure to deliver DNA into actinomycetes (Hopwood et al., 1985; Thompson et al., 1982). This technique allows the genetic manipulation of actinomycetes in order to obtain *papR2* overexpression strains by taking up pGM190/*papR2*. The method is based on the digestion of the cell wall and application of free DNA to the remaining protoplasts.

For generation of protoplasts, spores from glycerol stocks of the selected *Streptomyces* and *Kitasatospora* strains were cultivated 2–4 days in 50 mL S medium at 30 °C and 150–180 rpm (Tables 2.1, 2.5). The cell culture was transferred to 50 mL Falcon tubes and centrifuged at 5000 rpm for 10 min. The pellet was resuspended in 10 mL of P buffer with 1 mg/mL lysozyme (Tables 2.6, 2.9). The digestion of the cell wall was followed by microscopic analysis and was stopped when protoplasts were visible by addition of ice-cold P buffer. Protoplasts were filtered through sterile cotton and pelleted by centrifugation for 5 min at 3500 rpm. Protoplasts were diluted in 0.5–2 mL P buffer depending on amount. Aliquots of 200 µL were prepared and either frozen at −20 °C or directly used for transformation.

2.2.4.4 Protoplast Transformation

Transformation of protoplasts was performed using polyethylene glycol (PEG) (Bibb et al., 1978). In this study, pGM190-based constructs containing a kanamycin resistance cassette were transferred via protoplast transformation to actinomycetes strains. The plasmids were previously extracted from *E. coli* DH5α and the methylation deficient *E. coli* ET12567 to obtain methylated and unmethylated DNA, respectively, since many streptomycetes contain a methyl-specific restriction system of DNA (Flett et al., 1997) (Table 2.1).

For protoplast transformation, 20 µL of DNA (1–1.5 µg) (Table 2.3) was added to the approximately 180 µL protoplasts and instantly supplemented with 500 µL P buffer with 25% PEG (Table 2.6). The transformation mixture was immediately spread on R5 agar and incubated overnight at 28 °C. After overnight incubation, the plates were overlayed with NB soft agar containing 1/100 of appropriate antibiotics in order to obtain a 1/1000 dilution for the whole plate. Plates were incubated at 28 °C for all strains 5–15 days. Colonies were repeatedly transferred to new GYM and SFM plates containing appropriate antibiotics (Tables 2.5, 2.7).

Furthermore, to confirm regeneration of protoplasts, 10 µL protoplasts were mixed with 500 µL P buffer. Additionally, to differentiate between growth of protoplasts and mycelium, 10 µL protoplasts were mixed with 0.01% SDS in H_2O and vigorously vortexed to destroy all protoplasts. Both solutions were spread on R5 agar and cultivated at 28 °C for 2–3 days.

2.2.4.5   Conjugation of Actinomycetes

Conjugation was used as alternative method to transfer DNA to actinomycetes when protoplast transformation was not successful (Flett et al., 1997). In this study, pGM1190-based and pRM4-based constructs containing an apramycin resistance cassette were transferred via conjugation (Table 2.3). For this approach *E. coli* ET12567/pUZ8002 (Table 2.1) was transformed with a pGM1190- or pRM4-based construct according to 2.2.4. An *E. coli* overnight culture was inoculated at $OD_{600} = 0.05$–0.1 in 20 mL LB medium containing appropriate antibiotics and cultivated at 37 °C and 220 rpm until OD_{600} of 0.4–0.8. The cells were centrifuged at 5000 rpm for 5 min. The pellet was washed twice from antibiotics by resuspension in 10 mL LB medium and centrifugation at 5000 rpm for 5 min. Finally, the pellet was resuspended in 1–2 mL of LB medium. Meanwhile, 20 µL spores of the actinomycetes were mixed with 500 µL YT and heat shocked for 10 min at 50 °C. The spore suspension was mixed with 200 µL of *E. coli* cells and the mixture was spread on SFM plates. The plates contained either 10 mM MgCl$_2$ or 20 mM MgCl$_2$ + 20 mM CaCl$_2$, respectively, due to some conjugations being more effective with higher salt concentrations (Gomez-Escribano, pers. communication). The plates were incubated at 28 °C for *Streptomyces* strains and at 25 °C for *Kitasatospora* strains overnight. Then, the plates were overlayed with 1 mL H$_2$O containing 20 µL of fosfomycin and 25 µL of apramycin from antibiotic stocks. Plates were kept at 28 °C until exconjugants were visible. Single exconjugants were transferred to GYM and SFM plates containing appropriate antibiotics (Tables 2.5, 2.7).

2.2.5   Cultivation of Actinomycetes for Secondary Metabolite Production and Compound Extraction

To determine if actinomycetes expressing *papR2* might synthesize new compounds or produce an increased amount of a known compound, aqueous and organic compounds were extracted and subsequently tested for bioactivity. To cultivate actinomycetes for secondary metabolite production in liquid culture, a seed culture with 50 mL of R5 medium in baffled flasks was inoculated with a 1 cm^2 agar block from grown GYM or SFM agar plates (as described in Section 2.2.2.2). After 3 days of cultivation, 5 mL of the pre-culture was used to inoculate 45 mL R5 or OM media as main culture (Table 2.5). 2 mL samples were harvested after 24 h, 48 h, 72 h, and 96 h of cultivation for bioassays in order to determine the time point of optimal secondary metabolite production. The samples were centrifuged at

10000 rpm for 5 min and the supernatant was concentrated with a centrifugal evaporator (SP Genevac EZ-2, "aqueous" program) for approximately 2 h. The 4-fold concentrated supernatant samples containing all produced metabolites were subsequently used for bioassays.

For extraction of organic compounds, 20 mL of culture was harvested at the previously defined optimal production time point. 20 mL ethyl acetate (EtAc) was added to the culture and incubated 3–6 h at room temperature under constant vertical rotation. After centrifugation at 4000 rpm for 10 min, the organic phase was transferred to a new Falcon tube and completely dried with the centrifugal evaporator (SP Genevac EZ-2, "Low BP" program). The concentrated extracts were solved in 1 mL 50% methanol (MeOH), resulting in a 20-fold concentrated ethyl acetate extract sample.

Finally, ca. 20–25 mL culture remained from the original 50 mL, which were transferred to a 50 mL Falcon tube and centrifuged at 5000 rpm for 15 min. The supernatant was concentrated with the centrifugal evaporator (SP Genevac EZ-2, "aqueous" program) to 3–5 mL, resulting in a 5–7-fold concentrated supernatant sample. These supernatant samples were used for the repetition of bioassays or stored at −80 °C.

Additionally, obtained ethyl acetate extracts and supernatant samples were sent to the laboratory of Dr. Anja Schüffler at the "Institut für Biotechnologie und Wirkstoff-Forschung" (IBWF), Mainz in frame of a cooperation with BASF SE (Ludwigshafen, Germany) to conduct bioassays against plant-pathogenic fungi (*Botrytis cinerea, Fusarium culmorum,* and *Phytophthora infestans*) and insect larvae (*Plutella* and *Heliothis*). The data are summarized in Supplementary Tables S28–S41.

Supernatant samples of selected strains were also sent to the laboratory of Prof. Dr. Heike Brötz-Oesterhelt at the Institute of Microbiology and Infection Medicine of Eberhard Karls University of Tübingen for carrying out mode of action-specific analyses. Samples were tested against the bioreporter strains *B. subtilis yorB-lacZ, yppS-lacZ,* and *lialI-lacZ*, which had been shown to be specifically induced upon the following antibiotic stresses: *yorB* is induced by compounds interfering with DNA synthesis and structure, *yppS* is induced by RNA polymerase inhibitors, and *lialI* is induced by compounds causing cell envelope stress (Wex et al., 2021). In the bioreporter strains the respective promoters are coupled to the *lacZ* reporter system, thus reporter induction is detectable as a blue halo at the antibiotic diffusion borders (Wex et al., 2021). The bioreporter assays were performed by Anne Wochele (AG H. Brötz-Oesterhelt, University of Tübingen).

2.2.6 Bioactivity Assays

Concentrated supernatants and extracts from actinomycetes cultures were used for bioassays using different microorganisms as tests organisms including gram-positive and gram-negative bacteria as well as fungi (Table 2.2). To test for bioactivity, 100 µL of supernatant sample or ethyl acetate extract sample was pipetted into wells made in agar plates inoculated with the indication test strain. The plates were incubated according to the growth characteristics of the test strains. All bioassays were performed for at least 2 independent biological replicates. The degree of inhibition was assessed visually and for quantitative analysis the halo width from the border of the agar plug to the border of the inhibition zone was measured using ImageJ software (Schneider et al., 2012). Data analysis was performed with MS Excel and data are summarized in Supplementary Tables S11–S27 and Supplementary Table S42.

2.2.6.1 *E. coli* Bioassays

200–500 µL of an *E. coli* overnight culture was inoculated in 10 mL LB medium and incubated at 37 °C and 220 rpm until an OD_{600} of 0.4 according to Section 2.2.2.1. For one bioassay plate, 200 µL of *E. coli* culture were added to 20 mL of LB agar. For bioassays, the plates were incubated at 37 °C overnight (Tables 2.2, 2.5).

2.2.6.2 *M. luteus* Bioassays

A *M. luteus* culture was inoculated from an overnight culture and incubated at 30 °C and 220 rpm until an OD_{600} of 0.8. For one bioassay plate 400 µL of *M. luteus* culture was added to 20 mL of LB agar. For bioassays, the plates were incubated at 30 °C overnight (Tables 2.2, 2.5).

2.2.6.3 *Botrytis Cinerea* Bioassays

For *B. cinerea* bioassays, KIV agar plates were prepared and an agar block of a grown *B. cinerea* plate was placed in the center of the plate. The samples were positioned radially around the fungus and the plate was incubated for 14 days at room temperature in the dark (Tables 2.2, 2.5).

2.2.6.4 *Saccharomyces Cerevisiae* Bioassays

20 µL of a glycerol stock of *S. cerevisiae* was inoculated in 50 mL HM-medium and incubated for two days at 26 °C. For one bioassay plate, 250 µL of *S. cerevisiae* culture was added to 25 mL of HM agar. For bioassays, the plates were incubated at 26 °C for 2 days (Tables 2.2, 2.5).

2.2.7 High Performance Liquid Chromatography (HPLC) Analysis

Interesting bioactive samples and controls were analyzed by reverse-phase HPLC. Samples were centrifuged for 15 min at 9000 g and 4 °C and 500 µL of supernatant was applied to an HPLC vial. Samples (5–15 µl) were analyzed in an Agilent *1260 Infinity II* fitted with an *InfinityLab Poroshell 120 EC-C18* column (3.0 x 100 mm, 2.7-micron, 1,000 bar) and detected at 210 nm, 240 nm, 280 nm, 350 nm, and 450 nm. Samples were separated by 2 methods (the used method is stated in the chromatogram description):

1) Separation with a water-MeOH gradient from 1% to 100% for 20 min, at 0.75 µl/min flow. Buffer A was water with 0.1% of formic acid and Buffer B was 100% MeOH.
2) Separation with a water-acetonitrile (ACN) gradient from 5% to 100% for 20 min, at 0.5 µl/min flow. Buffer A was water with 0.1% of formic acid and Buffer B was 100% ACN.

2.2.8 Sodium Dodecyl Sulfate (SDS)—Polyacrylamide Gel Electrophoresis (PAGE)

To investigate the DNA binding capacity of modified complete PapR2 protein and a shortened PapR2 derivative, electrophoretic mobility shift assays (EMSAs) were performed. The purified Strep-tagged PapR2 proteins were kindly provided by Konrad Büssow (AG Wulf Blankenfeldt, HZI). In short, the native, as well as 3' 257-bp shortened *papR2* gene sequence have been synthesized as *E. coli* codon usage optimized sequences and cloned into pET28(+) vectors, resulting in the constructs pET28(+)_papR2 and pET28(+)_papR2_257, respectively (GenScript Biotech). The constructs were used for *papR2* and *papR2-257* expression in *E. coli* BL21 (Büssow, pers. communication).

To confirm the presence of the respective PapR2 proteins, SDS PAGE was performed. Separation of proteins was realized with a 12% acrylamide gel according to the method of Laemmli (Laemmli, 1970). The resolving and stacking gel were prepared according to Table 2.8. 30 µL of protein samples were mixed with 20 µL protein sample buffer (2X) and incubated at 100 °C for 5 min. 40 µL samples and 10 µL "Prestained protein ladder" (Table 2.10) were each loaded to the polymerized acrylamide gel and electrophoresis was conducted at 230 V for 2 h. The gel was stained with Coomassie blue for 20 min under constant shaking at 60 rpm. Subsequently, the gel was washed twice with H_2O for 10 min and destained overnight with 20% EtOH and 10% acetic acid under constant shaking at 60 rpm.

2.2.9   Electrophoretic Mobility Shift Assays (EMSAs)

EMSA was performed with the Cy5-labeled *papR1* promoter and purified PapR2 protein samples using a 2% agarose gel as described previously (Mast et al., 2015) (Table 2.8). The samples were prepared as follows:

- 1 µL of Cy5-labeled *papR1* promoter region (4 ng/µL)
- 5 µL of protein sample
- 2 µL loading buffer (without BPB)

In the first electrophoresis the received protein samples of purified PapR2 proteins were tested for binding to pro*papR1* for 60 min at 100 V. In the second electrophoresis the protein-DNA interaction was tested for specificity. For this purpose, 0.5 µL, 1 µL, and 4 µL of unlabeled *papR1* promoter (122 ng/µL) were each added for individual controls. An amplificate of the kanamycin resistance gene (*aphII*, 50 ng/µL) was used as control. Purple loading dye was applied to the first lane as indicator of running distance. The electrophoresis was conducted 40 min at 100 V. Fluorescent Cy5-labelled DNA was visualized with the Laser-Scanner „Typhoon Trio+ Variable Mode Imager" (GE Healthcare) with a pass filter of 670 nm and a laser at 633 nm.

Results

3

3.1 Construction of conjugatable, replicative pGM1190/*papR2* overexpression plasmids

The aim of the work was to activate antibiotic biosynthetic gene cluster expression of prioritized genome-sequenced actinomycetes with the help of SARP-type regulators. PapR2 is a SARP-type regulator and pathway-specific activator of pristinamycin biosynthesis in *S. pristinaespiralis*. To achieve biosynthetic gene cluster activation with the help of PapR2, suitable overexpression constructs are a prerequisite. Two *papR2* overexpression constructs, pRM4/*papR2* and pGM190/*papR2*, have been generated and validated in previous work (Krause, 2021; Mast et al., 2015). The pRM4/*papR2* plasmid is an integrative plasmid, where *papR2* transcription is under control of the constitutively active erythromycin promoter *ermE*p*. pRM4/*papR2* contains an apramycin resistance cassette (*aac(3)-IV*) for selection and an *oriT* for conjugation (Krause, 2021). The pGM190/*papR2* construct is a replicative plasmid, where *papR2* transcription is under control of an inducible thiostrepton promoter (*tipA*p). The plasmid harbors a kanamycin (*aphII*) and thiostreption (*tsr*) resistance cassette for selection, but lacks an origin of transfer (*oriT*) required for conjugation, which is why pGM190/*papR2* can only be transferred to actinomycetes via protoplast transformation (Mast et al., 2015). In previous work it has been observed that the biosynthetic cluster activation effect is stronger when using replicative pGM190/*papR2* construct. However, its broader application for many different actinomycetes is limited due to requirement of the *tipA* gene being present in the recipient strain either in the genome or

---

Supplementary Information The online version contains supplementary material available at https://doi.org/10.1007/978-3-658-44552-2_3.

O. Rössler, *SARP-Driven Activation of Antibiotic Biosynthetic Gene Clusters in Actinomycetes*, BestMasters, https://doi.org/10.1007/978-3-658-44552-2_3

in a mobilized plasmid to achieve full activation (Takano et al., 1995). Another limitation is the absence of an *oriT* and thus the necessity for protoplast transformation as genetic manipulation technique, to which many actinomycetes are not amenable to.

Therefore, the aim was to generate a replicative construct for *papR2* overexpression, which can be transferred to actinomycetes by conjugation. As suitable target plasmid, the pGM190-derivative pGM1190 was chosen for the cloning procedure, which contains the *tipA* promoter for induction of gene transcription and *oriT*, which is required for intergenic conjugation from *E. coli* to *Streptomyces* (Figure 3.1A; Muth, 2018). The *papR2* fragment was excised as ~ 1 kb *Eco*RI/*Nde*I-fragment from pGM190/*papR2* (Supplementary Figure S1 A) and cloned into the *Eco*RI/*Nde*I-linearized ~ 6,9 kb pGM1190 vector. This resulted in the plasmid pGM1190/*papR2-tipAp*, where *papR2* transcription is under the control of the thiostrepton-inducible *tipA* promoter (Figure 3.1B).

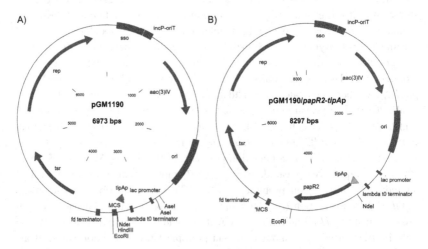

Figure 3.1 Plasmid maps of **A)** pGM1190 with *tipA* promoter, *oriT*, and apramycin resistance cassette *aac(3)-IV* and **B)** *papR2* overexpression construct pGM1190/*papR2-tipAp*

To overcome the limitation of *tipAp*, another pGM1190 construct with a constitutive *ermE* promoter (*ermEp**) was generated. For this purpose, *papR2-ermEp** was excised as ~ 1,5 kb *Ase*I/*Eco*RI fragment from pRM4/*papR2* (Supplementary Figure S1 B). The pGM1190 vector was digested with the same restriction enzymes, *Ase*I and *Eco*RI, removing the *lac* reporter system and

the *tipA* promoter. The linearized pGM1190 vector was ligated with *papR2-*
*ermE*p*, which resulted in the plasmid pGM1190/*papR2-ermE*p* (Figure 3.2),
where *papR2* transcription is under the control of constitutive *ermE* promoter.
Both plasmids, pGM1190/*papR2-tipA*p and pGM1190/*papR2-ermE*p* are self-
replicative plasmids and were used for conjugation of actinomycetes strains in
this study. The correctness of the constructs was verified by sequencing performed
by Eurofins Genomics GmbH.

Figure 3.2 Plasmid map
of *papR2* overexpression
construct pGM1190/
*papR2-ermE*p*

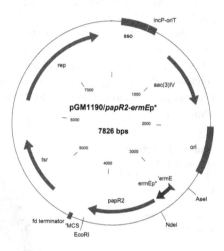

3.2 Verification of pGM1190-derived *papR2* overexpression constructs in regulator mutants of *S. coelicolor* A3(2)

In previous work it has been shown that PapR2 can complement for the regulatory
function of RedD in *Streptomyces lividans* (Krause et al., 2020). In *S. lividans*
and *S. coelicolor* A3(2) RedD is a SARP regulator essential for the production of
the red-pigmented secondary metabolite undecylprodigiosin (Red) that remains in
the mycelium. Deletion of *redD* in *S. coelicolor* M510 ($\Delta redD$) leads to loss of
Red production, (Floriano & Bibb, 1996), while production of the diffusible blue-
pigmented antibiotic actinorhodin (Act) is not affected (Figure 3.3). The double
mutant M512 ($\Delta redD$, ΔactII-ORF4) is unable to produce any of the pigmented
secondary metabolites, neither Red nor Act, thus no pigmentation of mycelium or

medium is expected. The mutant M513 carries a deletion of *afsR*, which encodes the large SARP-type regulator AfsR that acts as a pleiotropic regulator of secondary metabolism in *S. coelicolor* A3(2). AfsR was reported to affect Red and Act biosynthesis under certain growth conditions at a hierarchical higher level of regulation at which it cannot complement the lack of RedD or ActII-ORF4 (Floriano & Bibb, 1996). To test the functionality and strength of the generated *papR2* overexpression constructs pGM1190/*papR2-tip*Ap and pGM1190/*papR2-ermE*p* and the previously generated *papR2* overexpression constructs pGM190/*papR2* and pRM4/*papR2*, they were independently transferred to the well-studied regulatory mutants of the model organism *S. coelicolor* A3(2) and the effect on secondary metabolite biosynthesis was investigated. The respective empty vectors (pGM190, pRM4, and pGM1190) were transferred to the *S. coelicolor* mutants as controls. pGM190-based constructs were delivered via protoplast transformation, whereas pRM4- and pGM1190-based constructs were transferred via intergenic conjugation.

Figure 3.3 Actinorhodin production of *S. coelicolor* M510 (Δ*redD*) on R5 agar incubated for 2 weeks at 28 °C. The boxes highlight spots of secreted actinorhodin

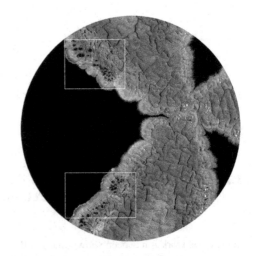

The previously validated constructs pGM190/*papR2* and pRM4/*papR2* were first transferred to *S. coelicolor* mutants. Production of Red was readily visible in most exconjugant colonies on the conjugation plates (Supplementary Figures S2 and S3). As R5 is a culture medium known to support pigmented-antibiotic production by *S. coelicolor* (Kieser et al., 2000), R5 agar plates were inoculated with the different *S. coelicolor* strains to assess production of secondary metabolites (Figure 3.4). In the case of the pGM190-derived transformants, addition of thiostrepton is required to fully induce the *tipA* promoter and thus *papR2* expression (Takano et al., 1995), but it is known that *tipA*p is active at low level without requirement for thiostrepton. Therefore, thiostrepton was not added in this study to not artificially influence compound production while expecting a low level of *papR2* expression sufficient for BGC activation. The phenotype of mutants containing pGM190 or pRM4 empty vectors was comparable to that of the respective mutant strain, showing that the empty vector does not influence the phenotype (Figure 3.4 A-F). The expression of *papR2* from both pGM190- and pRM4-based constructs completely restored Red production in M510 (Δ*redD*) to a similar extent to the production of the parental strain M145 (Figure 3.4 A, B). Overexpression of *papR2* from both pGM190/*papR2* and pRM4/*papR2* in the double mutant M512 (Δ*redD*, Δ*act*II-ORF4) restored Red but not Act production, which was expected as PapR2 should only complement the regulatory function of RedD but not ActII-ORF4 (Figure 3.4 C, D). Overexpression of *papR2* from both pGM190- and pRM4-based constructs in M513 (Δ*afsR*) led to enhanced production of Red and did not influence Act production (Figure 3.4 E, F). Furthermore, a PapR2-dependent delay of Act production was visible in M510 and M513 mutants containing pGM190/*papR2* and pRM4/*papR2* (Figure 3.4 A, B, E, F). Generally it was noticeable that Act production was delayed when Red was produced as the mutant strain M510 already produced Act in large amounts while the parental strain M145 only showed marginal Act production (Figure 3.4 A, B). These results confirmed the previous observation with *S. lividans* as PapR2 complemented for the regulatory function of RedD (Krause et al., 2020).

To test the functionality of the newly generated pGM1190-based *papR2* expression constructs they were transferred to the same regulatory mutant strains of *S. coelicolor*. pGM1190/*papR2-tipA*p did not provide clear activation of Red production in any of the *S. coelicolor* regulatory mutant strains. The results with pGM1190/*papR2-erm*Ep* however were very similar to those obtained with pGM190/*papR2* and pRM4/*papR2*, with production of Red observed already in colonies from M510 and M512 exconjugants on the conjugation plates and after re-striking on SFM (Supplementary Figures S2 and S3). The results obtained from *S. coelicolor* M510 and M512 with pGM190/*papR2* and pRM4/*papR2* cultivated on R5 agar were however difficult to reproduce with pGM1190/*papR2-erm*Ep*. *S. coelicolor* M510 and M512 pGM1190/*papR2-erm*Ep* did not show Red production on R5 agar, although the respective exconjugants previously produced Red on SFM agar. Several hypotheses are currently being investigated, e.g. a more readily loss of the construct when antibiotic selection is not maintained,

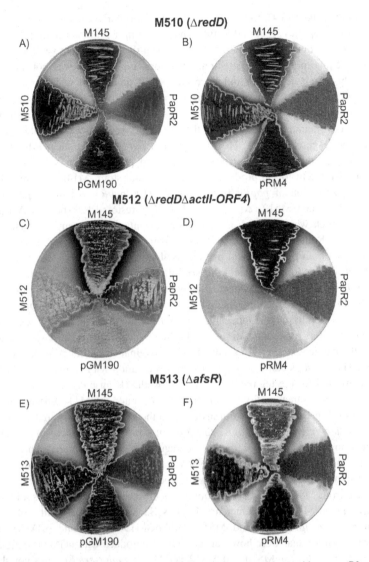

Figure 3.4 *S. coelicolor* parental strain M145 and mutants with or without *papR2* over-expression constructs on R5 agar. **A)** M510 (Δ*redD*) with pGM190/*papR2* ("PapR2") and pGM190, **B)** M510 (Δ*redD*) with pRM4/*papR2* ("PapR2") and pRM4, **C)** M512 (Δ*redD*, Δ*act*II-ORF4) with pGM190/*papR2* and pGM190, **D)** M512 (Δ*redD*, Δ*act*II-ORF4) with pRM4/*papR2* and pRM4, **E)** M513 (Δ*afsR*) with pGM190/*papR2* and pGM190, **F)** M513 (Δ*afsR*) with pRM4/*papR2* and pRM4. Pictures were taken after A) 48 h, B) 48 h, C) 120 h, D) 72 h, E) 96 h, and F) 72 h

since the R5 plates do not contain an antibiotic because the control strains M145, M510, M512 and M513 do not carry the corresponding resistance genes.

In summary, expression of *papR2* and functionality was confirmed for pGM190/*papR2* and pRM4/*papR2* with *S. coelicolor* mutants M510 (Δ*redD*), M512 (Δ*redD*, Δ*act*II-ORF4), and M513 (Δ*afsR*) as PapR2 complemented for the regulatory function of RedD. When PapR2 was expressed from pGM1190/ *papR2-ermEp** it also showed complementation for RedD function, which was however lost due to unknown reasons. On the contrary, for pGM1190/*papR2-tipAp* construct functionality was not confirmed as PapR2 was not reliably active. pGM190/*papR2* was further used for generation of *papR2* overexpression strains of prioritized actinomycetes. Regardless to the observation that no complementation was observed with the pGM1190/*papR2-tipAp* construct in the M510 and M512 mutants, pGM1190/*papR2-tipAp* as well as pRM4/*papR2* were used for cluster activation attempts with prioritized actinomycetes strains due to the inaccessibility of some strains to protoplast transformation (Section 3.1).

A BLASTP comparison of the PapR2 amino acid sequences with the respective three SARP-like regulatory proteins RedD, ActII-ORF, and AfsR from *S. coelicolor* A3(2) was conducted and demonstrated that PapR2 and RedD have the highest overall sequence similarity (44 % identity, 54 % similarity), whereas sequence similarities with ActII-ORF4 and AfsR were lower (Table 3.1). This was especially true for AfsR, which showed only partial sequence coverage with PapR2 (indicated by the gap percentage value of 8 %), which is due to the fact that AfsR belongs to the type of "large SARP regulators", whereas the other two regulators belong to the "small SARP type regulators". Altogether the experimental data show that PapR2 is not able to complement for the functionality of all SARP-type regulators. Thus, PapR2 may only activate BGCs with SARPs that are sufficiently similar to PapR2.

Table 3.1   BLASTP results from comparison of *S. coelicolor* A3(2) SARP regulatory proteins with PapR2

Protein compared to PapR2	Identity	Similarity	Gaps
RedD (Accession: AAA88556)	44 %	54 %	5 %
ActII-ORF (Accession: WP_003973893)	36 %	48 %	1 %
AfsR (Accession: WP_011029645)	40 %	51 %	8 %

3.3   Bioinformatic prioritization of strains for *papR2* expression

To identify DSMZ strains that carry secondary metabolite BGCs with a gene coding for a SARP transcriptional regulator, an advanced bioinformatic search for candidate genomes was conducted with all available genomes from the Integrated Microbial Genomes facility (IMG) of the Joint Genome Institute (JGI). The genomes were filtered by the order Streptomycetales and "DSM" as Genome Name/Sample name. 109 counts resulted from this search which were selected for further analysis. All selected genomes were analyzed for the occurrence of PapR2 homologs. For this purpose, a BLASTP search was conducted for all 109 counts using the PapR2 amino acid sequence (see supplementary material) with default parameters (e-value = $1e^{-5}$; 500 hits). The resulting candidate genes were filtered by a bit score > 175. In total, 33 genes were identified with a bit score > 175, from which five genes had a bit score > 300 (Table 3.2). Additionally, *S. ambofaciens* DSM 40697 with a bit score of 150 was included. Altogether, seven strains from the BLASTP search were selected for this study (Table 3.3). Furthermore, the 300 kb sequence surrounding the identified *papR2* homologous genes of the seven selected strains was analyzed with PatScanUI for occurrence of the PapR2 binding motif (GTCAGSSnnnnGTCAGSSnnnnGTCAGSS). Indeed, the binding motif (with 1–4 errors) occurred in all seven selected strains in intergenic regions, further suggesting a high probability of PapR2 having an effect on gene expression (Gomez-Escribano, pers. communication). Interestingly, the PapR2 binding motif occurred once without any errors within the genome of *S. leeuwenhoekii* (data not shown).

In addition to the mentioned strains, *Streptomyces* sp. TÜ4106, TÜ4128, and TÜ4134 were additionally involved in the study as *papR2* overexpression constructs have already been transferred to these strains in previous studies (Krause, 2021). *Streptomyces* sp. DSM 40976 was included because in ongoing lab experiments it has been found that this strain harbors seven SARP genes. In summary, eleven strains were used for studies on activation of BGCs using the PapR2 transcriptional activator.

Table 3.2 Candidate strains containing genes with amino acid sequences similar to PapR2 with bit score >175 and one strain with a bit score = 150

Strain	Bit score
Streptomyces leeuwenhoekii DSM 42122	390
Streptomyces sp. DSM 42143	389
Streptomyces resistomycificus DSM 40133	369
Kitasatospora cineracea DSM 44780	328
Kitasatospora niigatensis DSM 44781	327
Streptomyces griseoviridis DSM 40229	237
Streptomyces luteogriseus DSM 40483	199
Streptomyces corchorusii DSM 40340	197
Streptomyces vitaminophilus DSM 41686	196
Streptomyces cellostaticus DSM 40189	188
Streptomyces stelliscabiei DSM 41803	188
Kitasatospora viridis DSM 44826	187
Streptomyces cinnamonensis DSM 40803	185
Streptomyces costaricanus DSM 41827	184
Streptomyces griseochromogenes DSM 40499	184
Streptomyces corchorusii DSM 40340	182
Streptomyces leeuwenhoekii DSM 42122	182
Streptomyces auratus DSM 41897	182
Streptomyces misionensis DSM 40306	182
Streptomyces viridochromogenes DSM 40736	182
Kitasatospora atroaurantiaca DSM 41649	181
Streptomyces herbaricolor DSM 40123	180
Streptomyces bungoensis DSM 41781	179
Streptomyces melanosporofaciens DSM 40318	179
Streptomyces sp. DSM 41269	179
Streptomyces antibioticus DSM 40234	178
Streptomyces capillispiralis DSM 41695	178
Sreptomyces phaeogriseichromatogenes DSM 40710	178
Streptomyces platensis DSM 40041	178
Streptomyces avermitilis DSM 46492	177

(continued)

Table 3.2 (continued)

Strain	Bit score
Streptomyces caelestis DSM 40084	177
Streptomyces malaysiensis DSM 4137	176
Streptomyces stelliscabiei DSM 41803	176
Streptomyces ambofaciens DSM 40697	150

3.4   Generated *papR2* overexpression strains

To analyze *papR2* specific activation of BGCs, nine *papR2* overexpression strains were generated in the scope of this study, which are listed in Table 3.3. Additionally, three *papR2* overexpression strains generated in previous studies were involved in the study. The *papR2* overexpression strains and respective controls were cultivated in liquid culture for secondary metabolite production to perform bioassays with obtained supernatant samples and ethyl acetate extracts containing the organic compounds. *papR2* overexpression strains and respective controls (WT or the respective empty vector) were cultivated in R5 and Oat meal (OM) medium. Supernatant samples of 2 mL were taken from 24–96 h every 24 h hours to determine the time point of optimal bioactivity for each strain. All further supernatant samples and extracts were obtained at the determined time point of optimal bioactivity. Organic compounds were extracted with ethyl acetate and dried extracts were resuspended in 50 % MeOH. Supernatant samples and ethyl acetate extracts were each concentrated up to 4–7-fold and 20-fold concentration, respectively and subsequently used for bioassays. For bioassays, a panel of test strains was applied, which included different species of gram-positive (*M. luteus, B. subtilis*) and gram-negative bacteria (*E. coli* K12 and a set of outer membrane mutants more susceptible to antibiotics), and fungi (*S. cerevisiae* and *B. cinerea*) (Table 2.2). Bioactivity was compared between samples obtained from *papR2* overexpression strain and control strain based on the formation of inhibition halos. The inhibition halos were quantitatively assessed by measuring the width from the border of the well to the border of the halo. At least two independent biological replicates were analyzed for each strain. When stronger bioactivity of samples obtained from *papR2* overexpression strain was observed, at least three independent biological replicates were produced to confirm that the bioactivity was PapR2-specific. Selected supernatant samples and ethyl acetate extracts were analyzed with HPLC and fractionated for compound purification purposes.

Table 3.3   *papR2* overexpression strains with respective plasmids, which were generated in this study and in previous studies

Strains generated in this study	Plasmids
S. ambofaciens DSM 40697	pGM190; pGM190/*papR2*
S. avermitilis DSM 46492	pGM1190/*papR2-tip*Ap; pRM4/*papR2*
S. leeuwenhoekii DSM 42122	pGM190/*papR2*
S. leeuwenhoekii M1653	pGM1190; pGM1190/*papR2-tip*Ap
S. misionensis DSM 40306	pGM1190/*papR2-tip*Ap; pRM4/*papR2*
S. platensis DSM 40041	pGM190; pGM190/*papR2*
Streptomyces sp. DSM 40976	pGM1190/*papR2-tip*Ap
K. cineracea DSM 44780	pGM1190/*papR2-tip*Ap; pRM4/*papR2*
K. niigatensis DSM 44781	pGM1190/*papR2-tip*Ap; pRM4/*papR2*
Previously generated strains	**Plasmids**
Streptomyces sp. TÜ4106	pGM190; pGM190/*papR2*
Streptomyces sp. TÜ4128	pGM190; pGM190/*papR2*
Streptomyces sp. TÜ4134	pGM190; pGM190/*papR2*

3.4.1   Actinomycetes strains that did not show *papR2*-specific activity

Of the eleven actinomycetes strains that were selected for this study (Section 3.4), five strains did not show consistent PapR2 specific changes in bioactivity as outlined in detail below:

1) *Streptomyces avermitilis* DSM 46492 pRM4/*papR2* and pGM1190/*papR2-tip*Ap
2) *Streptomyces leeuwenhoekii* DSM 42122 pGM190/*papR2* and M1653 pGM1190/*papR2-tip*Ap
3) *Streptomyces misionensis* DSM 40306 pGM1190/*papR2-tip*Ap
4) *Streptomyces* sp. TÜ4134 pGM190/*papR2*
5) *Kitasatospora cineracea* DSM 44780 pRM4/*papR2* and pGM1190/*papR2-tip*Ap

Cultivation of *S. avermitilis* DSM 46492 WT and pGM1190/*papR2-tip*Ap in R5 led to change of medium coloration to dark brown while *S. avermitilis* pRM4/*papR2* culture displayed a yellowish medium color. *S. avermitilis* WT produces

melanin as reported by literature (Omura et al., 2001) and as indicated by two
recognized BGCs for melanin production in the antiSMASH analysis (Supplementary Figure S7). *S. avermitilis* WT, pRM4/*papR2*, and pGM1190/*papR2-tipA*p
displayed bioactivity against *Micrococcus luteus*, *E. coli* (Δ*lpxC*), and *Botrytis*
cinerea (Supplementary Table S12). The IBWF, which performs bioassays against
plant-pathogenic fungi and insect larvae, reported that extracts of *S. avermitilis*
DSM 46492 WT displayed bioactivity against *B. cinerea*, *Fusarium culmorum*,
and *Phytophthora infestans* (R5, GYM, and HM) and against *Plutella* larvae (R5
and GYM). *S. avermitilis* DSM 46492 pRM4/*papR2* only displayed bioactivity
against *F. culmorum* (R5 medium) and against *Plutella* larvae (GYM medium)
(Supplementary Table S29).

   S. leeuwenhoekii DSM 42122 WT and pGM190/*papR2* R5- and OM-derived
supernatants and extracts displayed strong bioactivity against all *E. coli* mutant
test strains, *M. luteus*, and *B. cinerea* (Supplementary Table S13). However, on
account of the production of the strong antimicrobial activity of chaxamycin,
it rendered to be difficult to assess differences in bioactivity between control and *papR2* overexpression strain samples of *S. leeuwenhoekii* DSM 42122.
Consequently, the chaxamycin non-producing mutant *S. leeuwenhoekii* M1653
pGM1190 and pGM1190/*papR2-tipA*p strains were generated and used for bioactivity studies. *S. leeuwenhoekii* M1653 pGM1190 and pGM1190/*papR2-tipA*p
strains displayed bioactivity against *E. coli* (Δ*tolC*), (Δ*lpxC*), *M. luteus*, *B.*
subtilis, (Supplementary Table S14) and *Plutella* larvae (Supplementary Tables
S30 and S31). The R5-derived supernatant samples of *S. leeuwenhoekii* M1653
pGM1190 and pGM1190/*papR2-tipA*p were also sent to the University of Tübingen to analyze the mode of action of bioactive compounds with bioreporter strains
of *B. subtilis*. No specific induction of *lacZ* expression was observed in bioreporter assays. The blue halo directly at the sample visible in all bioreporter
assays was likely caused by the intrinsic β-galactosidase (H. Brötz-Oesterhelt,
pers. communication).

   S. misionensis DSM 40306 pGM1190/*papR2-tipA*p and WT displayed bioactivity against gram-positive *M. luteus* and the fungi *B. cinerea* and *S. cerevisiae*
(Supplementary Table S15). Furthermore, *S. misionensis* DSM 40306 pRM4/
papR2 and WT exhibited bioactivity against IBWF test strains *B. cinerea*, *F.*
colmorum, *P. infestans*, and *Plutella* larvae (Supplementary Table S32).

   Streptomyces sp. TÜ4134 pGM190 and pGM190/*papR2* produced a bright
orange pigment during cultivation in R5 medium and on GYM agar. R5 medium
coloration of *Streptomyces* sp. TÜ4134 pGM190/*papR2* appeared in a darker
orange than the culture of *Streptomyces* sp. TÜ4134 pGM190. *Streptomyces* sp.
TÜ4134 pGM190 and pGM190/*papR2* supernatant samples exhibited equally

strong bioactivity against all *E. coli* mutant test strains and *M. luteus*, whereby extract samples exhibited bioactivity only against *E. coli* ($\Delta tolC$) (Supplementary Table S23). *Streptomyces* sp. TÜ4134 WT extract samples displayed bioactivity against *B. cinerea*, *F. culmorum*, and *P. infestans* (Supplementary Table S39). *Streptomyces* sp. TÜ4134 has not been genome sequenced but was reported as producer of the antibiotics questiomycin A and zinc-coproporphyrin III (data from Tübingen Stammsammlung, Mast, pers. communication). Hence, potential differences in bioactivity between *Streptomyces* sp. TÜ4134 WT pGM190/*papR2* and pGM190 may have not been detected because of questiomycin production or due to the absence of PapR2 homologous SARPs.

K. cineracea DSM 44780 WT, pRM4/*papR2*, and pGM1190/*papR2-tipA*p produced a pink pigment during cultivation in R5. Cultures of *K. cineracea* pRM4/ *papR2* and pGM1190/*papR2-tipA*p appeared in a darker medium coloration compared to the WT. Bioactivity induced by *K. cineracea* DSM 44780 WT, pRM4/ *papR2*, and pGM1190/*papR2-tipA*p was generally weak and rarely reproducible (Supplementary Table S25). OM-derived supernatants displayed very low inhibition of *E. coli* ($\Delta tolC$). R5-derived supernatants displayed low bioactivity specific for *K. cineracea* pRM4/*papR2* and pGM1190/*papR2-tipA*p against *E. coli* ($\Delta tolC$), ($\Delta lpxC$), and *M. luteus* while the WT did not show bioactivity. Furthermore, the IBWF reported bioactivity against *Plutella* larvae from *K. cineracea* DSM 44780 WT and pRM4/*papR2* extracts (Supplementary Table S40).

In summary, the majority of strains where no *papR2* specific bioactivity was observed contained the pGM1190/*papR2-tipA*p construct, which was also not functional in the *S. coelicolor* mutant strains. Thus, the inability of BGC activation indicated by stronger bioactivity might be due to lack of *papR2* expression. These strains are still potentially good candidates for *papR2* specific activation of BGCs according to bioinformatic analyses and should be envisaged for further BGC attempts with alternative *papR2* overexpression constructs.

3.4.2   *Streptomyces* sp. TÜ4128

To test the effect of *papR2* expression on secondary metabolite production, *Streptomyces* sp. TÜ4128 pGM190/*papR2* and the control strain *Streptomyces* sp. TÜ4128 pGM190 were cultivated in R5 and OM media as five independent biological replicates. The time point of maximum bioactivity against test strains was determined between 72 and 96 h of cultivation. Cultivation of *Streptomyces* sp. TÜ4128 pGM190/*papR2* in R5 medium caused the production of a green-blue pigment, which was mostly absent in the *Streptomyces* sp. TÜ4128 pGM190

control culture (Figure 3.5A). Additionally, cultivation of both *Streptomyces* sp.
TÜ4128 pGM190/*papR2* and pGM190 in OM medium caused the production
of a red-brown pigment, whereby TÜ4128 pGM190/*papR2* appeared intensively
darker in medium coloration (Figure 3.5B).

Figure 3.5 *Streptomyces*
sp. TÜ4128 pGM190 (left)
and pGM190/*papR2* (right)
cultivated 72 h in **A)** R5
and **B)** OM media

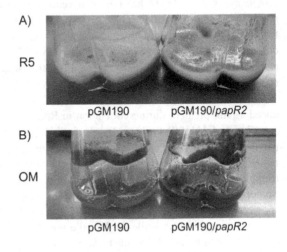

R5-derived supernatants exhibited weak and inconsistent bioactivity against
gram-negative *E. coli* (Δ*lpxC*) with no difference between TÜ4128 pGM190/
papR2 and pGM190 samples. R5- and OM-derived supernatants samples of
TÜ4128 pGM190/*papR2* and pGM190 also exhibited weak and inconsistent
activity against both gram-positive bacterial test strains *M. luteus* and *B. subtilis*.
Generally, a consistently and significantly stronger bioactivity for supernatant
samples of TÜ4128 pGM190/*papR2* was not detected (Supplementary Table
S21). R5-derived supernatant samples of TÜ4128 pGM190/*papR2* and pGM190
strains were also used for bioassays against *B. subtilis lacZ* reporter strains, where
no specific induction of *lacZ* expression was observed (H. Brötz-Oesterhelt, pers.
communication) and thus, no mode of action for the bioactive compound can be
proposed.

R5-derived ethyl acetate extracts of TÜ4128 pGM190/*papR2* exhibited bioactivity against *M. luteus* while the respective control samples of TÜ4128 pGM190
never showed any bioactivity against *M. luteus* (Figure 3.6). Also OM-derived
ethyl acetate extracts of TÜ4128 pGM190/*papR2* displayed stronger bioactivity
than the respective control samples of TÜ4128 pGM190, however, bioactivity
from OM-derived extracts was less frequently observed than from R5-derived

extracts. R5- and OM-derived ethyl acetate extracts of both TÜ4128 pGM190/ *papR2* and pGM190 strains displayed consistently bioactivity against *B. subtilis* with no significant differences (Supplementary Table S22). Antifungal activity was neither detected in this study nor by the IBWF (Supplementary Table S38). The width of inhibition halos was not measured for *Streptomyces* sp. TÜ4128 pGM190/*papR2* and pGM190 strains due to the inconsistency of bioactivity. There was however a consistent production of a green-blue pigment in TÜ4128 pGM190/*papR2* when cultivated in R5, which was not seen in the TÜ4128 pGM190 control or at a much lower level, indicating that PapR2 induces a biosynthetic pathway. It is also worth pointing out that bioactivity of ethyl acetate extracts against *M. luteus* did not correlate with pigment production.

Figure 3.6 Bioassay results of *Streptomyces* sp. TÜ4128 control strain (pGM190, "V") and *papR2* overexpression strain (pGM190/*papR2*, "PapR2"). 20-fold ethyl acetate extracts ("EtAc") from cultures cultivated 96 h in R5 and OM were tested against *M. luteus*

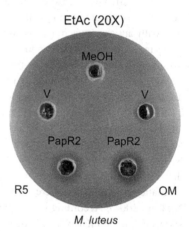

EtAc (20X)

M. luteus

AntiSMASH predicted 46 BGCs for the genome of TÜ4128, of which five contain SARP genes (Supplementary Figure S42). Overall, *Streptomyces* sp. TÜ4128 was the strain with the largest number of BGCs. The five BGCs containing SARP genes were predicted to belong to the following BGC types: NRPS (region 13.1), melanin (region 16.3), type-II PKS (region 39.1), butyrolactone (region 52.1), and phenazine (region 53.1). Region 16.3 was identified to potentially code for bagremycin, an antibiotic substance which has previously been reported as biosynthetic product of *Streptomyces* sp. TÜ4128 (Ye et al., 2019). The bagremycin BGC contains the SARP gene *bagI*, of which the gene product shows 51 % similarity to PapR2 (Supplementary Table S8). However, the bagremycin BGC is probably not the BGC activated by PapR2 in *Streptomyces* sp. TÜ4128, since bagremycin is not a green-blue pigment and the observed bioactivity in the current study is inconsistent with the reported bioactivity of bagremycin as *S. cerevisiae* and *B. cinerea* were inhibited while *M. luteus* was not (Bertasso et al., 2001).

Region 39.1 coding for a type-II PKS contains two SARP genes, of which the gene products show 57 % and 65 % similarity to PapR2 (Supplementary Table S8, Locus tags "ctg39_25" and "ctg39_35"). AntiSMASH predicted fluostatins M–Q as most similar BGC, however, cluster comparison with data deposited in the MiBIG database suggested that the BGC is most similar to the griseusin BGC of *Streptomyces griseus* (Yu et al., 1994). In a previous study, it has been shown that the griseusin BGC in *Streptomyces* sp. CA-256286 is activated upon heterologous expression of different SARP proteins including RedD, ActII-ORF4, and PapR2, whereby ActII-ORF4 was identified as the essential activator (Beck et al., 2021). Griseusin is an orange pigmented type pyranonaphthoquinones type II polyketide with anticancer activity, as well as antibiotic activity against gram-positive bacteria, such as *B. subtilis* and *S. aureus* (Tsuji et al., 1976). The bioactivity of griseusin against gram-positive bacteria would be consistent with the bioactivity against *M. luteus* and *B. subtilis* observed in this study, however, production of an orange pigment was not observed.

Region 53.1 contains a SARP gene, where the gene product shows the highest similarity to PapR2 with 58 % identity and 74 % similarity scores, although the amino acid length of PapR2 is twice as long as that of the unknown TÜ4128 SARP (Supplementary Table S8, Locus tag "ctg53_1"). Region 53.1 was predicted to resemble a phenazine BGC. Phenazines are often pigmented due to their aromatic structures and were observed to have antibiotic activities (Laursen & Nielsen, 2004). According to comparisons with the MiBIG database, region 53.1 shows certain similarity to a pyocyanine BGC, which is a blue pigmented natural compound from *Pseudomonas aeruginosa* PAO (Mavrodi et al., 2001). Pyocyanine was reported to exert bioactivity especially against gram-positive bacteria, of which *M. luteus* was the most susceptible to the antibiotic, and gram-negative bacteria such as *E. coli* were moderately susceptible, which was also observed in the current study (Baron & Rowe, 1981). Therefore, it is conceivable that PapR2 has activated the BGC region 53.1, which led to the blue pigmentation in the R5 culture.

3.4.3   *Streptomyces* sp. DSM 40976

To test the effect of *papR2* expression on secondary metabolite production, *Streptomyces* sp. DSM 40976 pGM1190/*papR2-tipA*p and *Streptomyces* sp. DSM

40976 WT as control were cultivated in R5 and OM media as three independent biological replicates. The time point of maximum bioactivity against test strains was determined between 48 h and 72 h of cultivation. Cultivation of *Streptomyces* sp. DSM 40976 pGM1190/*papR2-tipA*p and WT did not cause any change of medium coloration of R5 or OM.

Supernatant samples obtained from *Streptomyces* sp. DSM 40976 pGM190/ *papR2* and WT cultivated in R5 medium exhibited bioactivity against all *E. coli* mutant test strains and *M. luteus* with no difference between the *papR2* overexpression strain and WT (Supplementary Table S18). Supernatant samples obtained from *Streptomyces* sp. DSM 40976 pGM1190/*papR2-tipA*p and WT cultivated in OM medium only displayed bioactivity against *M. luteus*, whereby bioactivity was distinctly stronger from *papR2* overexpression strain than WT (Figure 3.7A). Measurements of the width of inhibition halos indicate an average width of 2.96 ± 1.9 mm for *Streptomyces* sp. DSM 40976 WT and 6.29 ± 3.02 mm for *Streptomyces* sp. DSM 40976 pGM1190/*papR2-tipA*p (Figure 3.7B), which clearly shows a stronger bioactivity specific for *papR2* expression.

R5- and OM-derived ethyl acetate extracts only exhibited bioactivity against *M. luteus* with no difference between *Streptomyces* sp. DSM 40976 pGM1190/ *papR2-tipA*p and *Streptomyces* sp. DSM 40976 WT (Supplementary Table S18). The IBWF detected more bioactivity of OM-derived extracts from *Streptomyces* sp. DSM 40976 pGM1190/*papR2-tipA*p against *Plutella* larvae compared to the *Streptomyces* sp. DSM 40976 WT (Supplementary Table S35). Antifungal activity from *Streptomyces* sp. DSM 40976 pGM1190/*papR2-tipA*p and WT was not observed.

AntiSMASH analysis predicted 32 BGCs for the genome of *Streptomyes* sp. DSM 40976, of which seven BGCs comprise nine SARP genes (Supplementary Figure S31). Overall, *Streptomyces* sp. DSM 40976 was the strain with the largest number of SARP genes located in identified BGCs. The seven BGCs belong to the following cluster types: type-I and type-II PKS, NRPS, lanthipeptides, and others. The gene products of five SARP genes have an amino acid similarity score of $\geq 50\%$ compared to PapR2, of which the highest score is 58 % for the SARP, where the coding gene is located in region 45.1 encoding a hybrid BGC, consisting of NRPS, NRPS-like, and type-I PKS core genes (Supplementary Table S6, Locus Tag: "ctg45_9"). Due to multiple opportunities for candidate BGCs that comprise more than one possible product, a suggestion for a substance cannot be made.

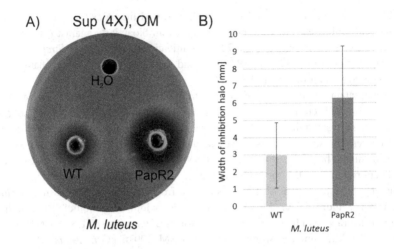

Figure 3.7 Bioassay results of *Streptomyces* sp. DSM 40976 WT and *papR2* overexpression strain (pGM1190/*papR2-tip*Ap, "PapR2") after 72 h cultivation in OM. **A)** Supernatant ("Sup", 4X-concentrated) was tested against *M. luteus*, with H_2O as negative control. **B)** Average width of inhibition halos against *M. luteus* (N = 3). Analysis was conducted with ImageJ and black lines show the standard deviation. N = Number of replicates

3.4.4   *Kitasatospora niigatensis* DSM 44781

To test the effect of *papR2* expression on secondary metabolite production, *Kitasatospora niigatensis* DSM 44781 pGM1190/*papR2-tip*Ap, pRM4/*papR2* and WT as control were cultivated in R5 and OM media as three independent biological replicates. The time point of maximum bioactivity against test strains was determined at 72 h of cultivation. Cultivation of *K. niigatensis* DSM 44781 WT and pGM1190/*papR2-tip*Ap in R5 medium caused a change of medium coloration from yellowish to pink-red, while cultivation of *K. niigatensis* DSM 44781 pRM4/*papR2* in OM medium did not cause a change of medium coloration (Figure 3.8). R5-derived supernatant samples obtained from *K. niigatensis* DSM 44781 pGM1190/*papR2-tip*Ap, pRM4/*papR2* and WT equally displayed bioactivity against gram-negative *E. coli* (Δ*tolC*) (Supplementary Table S26). OM-derived supernatant samples obtained from *K. niigatensis* DSM 44781 pGM1190/*papR2-tip*Ap and pRM4/*papR2* showed stronger bioactivity than supernatant samples obtained from *K. niigatensis* DSM 44781 WT (Figure 3.9A). For further analyses, only *K. niigatensis* DSM 44781 pGM1190/*papR2-tip*Ap and WT were used.

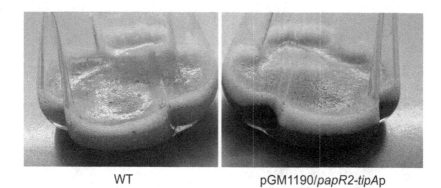

WT   pGM1190/*papR2-tip*Ap

Figure 3.8 *K. niigatensis* DSM 44781 WT (left) and pGM1190/*papR2-tip*Ap (right) cultivated 72 h in R5 medium

The average widths of inhibition halos confirmed a stronger inhibition from supernatant samples of *K. niigatensis* DSM 44781 pGM1190/*papR2-tip*Ap compared to *K. niigatensis* DSM 44781 WT with 2.63 ± 0.38 mm and 0.53 ± 0.75 mm, respectively (Figure 3.9B). Furthermore, ethyl acetate extracts of *K. niigatensis* DSM 44781 pRM4/*papR2* and WT cultivated 48 h in GYM and HM inhibited *Plutella* larvae as reported by the IBWF (Supplementary Table S41).

AntiSMASH analysis predicted 27 BGCs for the genome of *K. niigatensis* DSM 44781, of which five BGCs comprise seven genes coding for SARPs (Supplementary Figure S54). The candidate BGCs code for NRPS, type-II PKS, ladderanes and lanthipeptide-type BGCs. Phylogenomic analysis with the tool TYGS revealed that the strains *K. niigatensis* DSM 44781 and *K. cineracea* DSM44780 are highly similar to each other (data not shown; Mast pers. commun.). Due to the high phylogenetic similarity of both strains, all five BGCs are also found in *K. cineracea* DSM 44780 either in similar direction or inverted (Supplementary Figure S48). BGC 5.1 in *K. niigatensis* DSM 44781 is similar to BGC 4.1 in *K. cineracea* DSM44780, which comprises three SARP genes of which one displayed the highest amino acid similarity score with 81 % and 0 % gap percentage compared to PapR2 and thus might be the candidate cluster, which has been activated upon *papR2* expression (Supplementary Table S10, Locus tag: "EDD38_RS37695"). This BGC is a hybrid BGC consisting of NRPS, type-II PKS, butyrolactone etc. core genes. At this stage, it is not possible to predict a possible BGC product. Furthermore, for future analyses only one of the two *Kitasatospora* strains (DSM 44780 or DSM 44781) should be considered due to the high phylogenetic similarity.

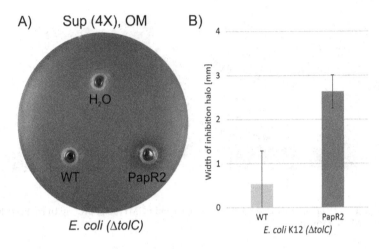

Figure 3.9   Bioassay results of *K. niigatensis* DSM 44781 WT and *papR2* overexpressing strain (pGM1190/*papR2-tipA*p, "PapR2") after 72 h cultivation in OM. **A)** 4-fold concentrated supernatant ("Sup (4X)") was tested against *E. coli (ΔtolC)*, with H_2O as negative control. **B)** Average width of inhibition halos against *E. coli (ΔtolC)* (N = 3). Analysis was conducted with ImageJ and black lines show the standard deviation. N = Number of replicates

3.4.5   *Streptomyces platensis* DSM 40041

To test the effect of *papR2* expression on secondary metabolite production, *S. platensis* DSM40041 pGM190/*papR2* and the control strain *S. platensis* DSM 40041 pGM190 were cultivated in R5 medium as three independent biological replicates and in OM medium as four independent biological replicates. The time point of maximum bioactivity was determined at 72 h of cultivation. *S. platensis* DSM 40041 pGM190/*papR2* consistently caused a change of OM medium coloration to orange, while the control retained the yellowish OM medium color (Figure 3.10).

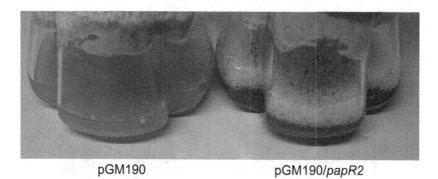

pGM190   pGM190/*papR2*

Figure 3.10   *S. platensis* DSM 40041 pGM190 (left) and pGM190/*papR2* (right) cultivated 72 h in OM medium

S. platensis DSM40041 pGM190/*papR2* and pGM190 R5-derived supernatants and extracts exhibited general bioactivity against all gram-negative bacterial test strains, as well as against *M. luteus* with no differences between control and *papR2* overexpression samples (Supplementary Tables S16 and S17). OM-derived supernatants and ethyl acetate extracts of *S. platensis* DSM40041 pGM190/*papR2* consistently exerted stronger bioactivity against all gram-negative (Figure 3.11) and all gram-positive bacterial test strains (Figure 3.12). In particular, super-natants and extracts obtained from *S. platensis* DSM40041 pGM190/*papR2* cultivated in OM medium consistently inhibited *E. coli* mutant strains Δ*tolB* (Figure 3.11A), Δ*tolC* (Figure 3.11B), Δ*lpxC* (Figure 3.11C), and the parental strain *E. coli* K12 (Figure 3.11D), more than the respective control supernatants and extracts of *S. platensis* DSM40041 pGM190. OM-derived supernatants of *S. platensis* DSM40041 pGM190/*papR2* inhibited the gram-positive bacte-ria *M. luteus* and *B. subtilis* stronger than the control samples of *S. platensis* DSM40041 pGM190 (Figure 3.12A). Additionally, OM-derived ethyl acetate extracts of *S. platensis* DSM40041 pGM190/*papR2* inhibited *M. luteus* but not *B. subtilis* stronger than the control extracts of *S. platensis* DSM40041 pGM190 (Figure 3.12B). Antifungal activity against *B. cinerea* was observed for OM-derived supernatant in this study (Supplementary Table S16) and by the IBWF (Supplementary Table S34). Additionally, the IBWF detected strong bioactivity of ethyl acetate extracts obtained from *S. platensis* DSM40041 pRM4/*papR2* cul-tivated 96 h in NL19 medium against *B. cinerea, F. culmorum,* and *P. infestans* while the respective control sample of *S. platensis* DSM40041 WT did not display any bioactivity (Supplementary Table S33).

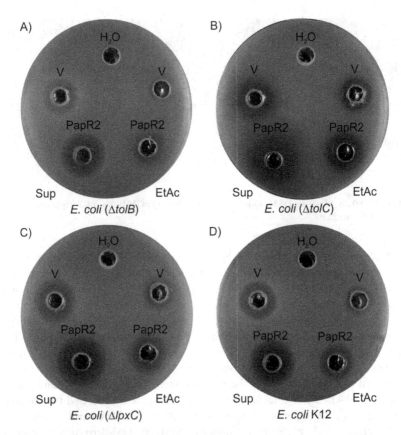

Figure 3.11 Bioassay results of *S. platensis* DSM 40041 control strain (pGM190, "V") and *papR2* overexpression strain (pGM190/*papR2*, "PapR2") after 72 h cultivation in OM medium. 7-fold concentrated supernatants ("Sup", left) and 20-fold concentrated extracts ("EtAc", right) were tested against **A)** *E. coli* (Δ*tolB*), **B)** *E. coli* (Δ*tolC*), **C)** *E. coli* (Δ*lpxC*), and **D)** parental *E. coli* K12

The average width of inhibition halos from supernatant samples of *S. platensis* DSM 40041 pGM190/*papR2* and pGM190 indicates a consistently increased bioactivity specific for *papR2* expression against *E. coli* strains Δ*tolB* and Δ*lpxC*, *M. luteus* and *B. subtilis* (Figure 3.13A). Inhibition halos of *E. coli* strains K12 and Δ*tolC* were not assessed because only two replicates were analyzed. Likewise,

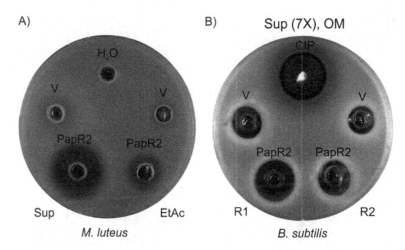

Figure 3.12 Bioassay results of DSM 40041 control strain (pGM190, "V") and *papR2* overexpression strain (pGM190/*papR2*, "PapR2") after 72 h cultivation in OM medium. **A)** 7-fold concentrated supernatant ("Sup", left) and 20-fold concentrated extracts ("EtAc", right) were tested against *M. luteus*. **B)** 7-fold concentrated supernatant samples of two independent replicates ("R1" and "R2") were tested against *B. subtilis*. Ciprofloxacin (CIP) was used as positive control

the average width of inhibition halos from ethyl acetate extracts of *S. platensis* DSM 40041 pGM190/*papR2* and pGM190 indicates a consistently increased bioactivity specific for *papR2* expression against *E. coli* strains Δ*tolB*, Δ*tolC*, and Δ*lpxC* as well as against *M. luteus* (Figure 3.13B).

To identify the secondary metabolite with bioactivity against *M. luteus*, the ethyl acetate extracts of three replicates of *S. platensis* DSM40041 pGM190/ *papR2* and as comparison from *S. platensis* DSM 40041 pGM190 were analyzed by HPLC. The HPLC analysis revealed multiple peaks at an absorbance of 280 nm which were significantly increased for *S. platensis* DSM40041 pGM190/*papR2* (Figure 3.14A-C) compared to *S. platensis* DSM40041 pGM190 (Figure 3.14D). To identify the bioactive compound, a preparative HPLC was conducted and the sample from 0.5–19.5 min was separated every minute into distinct fractions. Unfortunately, a distinct fraction with significant abundant bioactivity was not obtained since almost all fractions showed a low level of bioactivity.

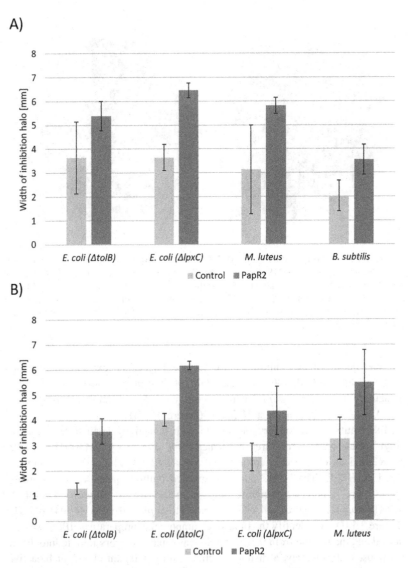

Figure 3.13 Average width of inhibition halos of OM-derived **A)** 7-fold concentrated supernatants against *E. coli* (Δ*tolB*) (N = 4), *E. coli* (Δ*lpxC*) (N = 3), *M. luteus* (N = 4), and *B. subtilis* (N = 3) and **B)** 20-fold concentrated ethyl acetate extracts against *E. coli* (Δ*tolB*) (N = 3), *E. coli* (Δ*tolC*) (N = 3), *E. coli* (Δ*lpxC*) (N = 4), and *M. luteus* (N = 4), after 72 h cultivation. Analysis was conducted with ImageJ and black lines show the standard deviation. N = Number of replicates

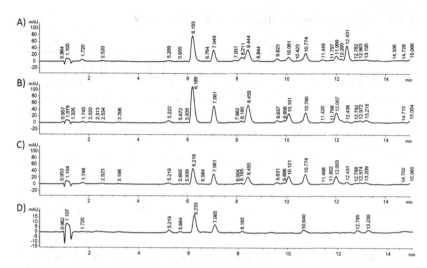

Figure 3.14 Comparative HPLC chromatogram of OM-derived ethyl acetate extracts of *S. platensis* DSM 40041 pGM190/*papR2* **A)** replicate 1, **B)** replicate 2, **C)** replicate 3, and **D)** *S. platensis* DSM 40041 pGM190 separated with H_2O-ACN gradient. Absorbance was measured at 280 nm

AntiSMASH analysis with the *S. platensis* genome sequence revealed 35 BGCs in total (Supplementary Figure S23). *S. platensis* is a known producer of several antibiotically active substances, including oxytetracycline, platencin, platensimycin, migrastatin, and several others (Ju et al., 2005; Singh et al., 2006). *S. platensis* is also a producer of antifungals including clavamycin A, dorrigocin A, and resormycin, whereby the latter is also a herbicide (Diana & Cirrincione; Diana & Cirrincione, 2015; Igarashi et al., 1997; Lim et al., 2009). The detection of generally high bactericidal activity in this study as well as antifungal activity detected by the IBWF and is consistent with *S. platensis* being a potent producer of several bioactive compounds and being known as general bioprotectant. Seven BGCs were found to contain in total eight genes coding for SARPs, including region 30 that is responsible for biosynthesis of oxytetracycline. The majority of amino acid sequences of putative SARP gene products indicate similarity scores between 48 % and 51 % compared to the PapR2 amino acid sequence (Supplementary Table S5). The putative SARP with the highest similarity of 58 % to PapR2 is located in region 31 (Locus tag "CP981_35175"), which was predicted to code for an unknown NRPS-like substance. Moreover, regions 30 and 31 were found to contain the SARP consensus sequence indicating a high probability for activation by PapR2.

3.4.6   *Streptomyces ambofaciens* DSM 40697

To test the effect of *papR2* expression on secondary metabolite production, *S. ambofaciens* DSM 40697 pGM190/*papR2* and the control strain *S. ambofaciens* DSM 40697 pGM190 were cultivated in R5 and OM media as three independent biological replicates. The time point of maximum bioactivity against test strains was determined at 72 h of cultivation. Both *S. ambofaciens* DSM 40697 strains did not cause a change of medium coloration in R5 and OM during cultivation.

Supernatant samples from *S. ambofaciens* DSM 40697 pGM190/*papR2* and pGM190 exhibited general bioactivity against numerous test strains including gram-positive and gram-negative bacteria as well as fungi, whereby bioactivity was especially prominent for supernatant samples obtained from R5 cultivation (Supplementary Table S11). In particular, R5-derived supernatant samples consistently exhibited bioactivity against all *E. coli* mutant strains, *M. luteus*, and *B. cinerea*, however the degree of inhibition varied dramatically in both *S. ambofaciens* pGM190/*papR2* and pGM190. R5-derived supernatants of *S. ambofaciens* pGM190/*papR2* and pGM190 were tested against *B. subtilis* reporter strains, however no specific induction of *lacZ* expression was observed (H. Brötz-Oesterhelt, pers. communication). OM-derived supernatant samples inconsistently displayed bioactivity against *E. coli* ($\Delta tolC$), *M. luteus*, and *B. cinerea* with no difference between *S. ambofaciens* pGM190/*papR2* and pGM190. Ethyl acetate extracts from *S. ambofaciens* pGM190/*papR2* and pGM190 cultivated in R5 displayed *papR2* specific bioactivity against the gram-positive bacteria *M. luteus* (Figure 3.15A) and *B. subtilis* (Supplementary Table S11). Bioactivity from R5- and OM-derived extracts against other test strains was not observed.

Measurements of the width of the inhibition halo against *M. luteus* showed an average of 1.80 ± 2.55 mm for *S. ambofaciens* pGM190 and 3.28 ± 1.23 mm for *S. ambofaciens* pGM190/*papR2* indicating a trend of stronger bioactivity specific for *papR2* expression, however with a high variability as highlighted by the standard deviation (Figure 3.15B). Furthermore, inhibition halos against *B. subtilis* displayed an average width of 1.49 ± 1.08 mm for *S. ambofaciens* pGM190 and 3.84 ± 0.59 mm for *S. ambofaciens* pGM190/*papR2*, demonstrating clearly stronger activity specific for *papR2* overexpression with less variability (Figure 3.15B). To summarize, a *papR2* specific stronger bioactivity against gram-positive bacteria was observed from ethyl acetate extracts obtained from *S. ambofaciens* pGM190/*papR2* cultivated in R5 medium, whereby inconsistent bioacitivities indicate labile production of the activated secondary metabolite.

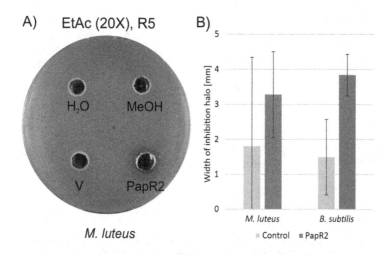

Figure 3.15 Bioassay results of *S. ambofaciens* DSM 40697 control strain (pGM190, "V") and *papR2* overexpression strain (pGM190/*papR2*, "PapR2") 20-fold concentrated ethyl acetate extracts after 72 h cultivation in R5. **A)** Bioassay against *M. luteus*, with H_2O and 50 % MeOH as negative controls. **B)** Average width of inhibition halos against *M. luteus* and *B. subtilis* (each N = 3). Measurements were conducted with ImageJ and black lines show the standard deviation. N = Number of replicates

To identify the secondary metabolite with bioactivity against gram-positive bacteria, the R5-derived ethyl acetate extracts of *S. ambofaciens* pGM190/*papR2* and pGM190 were comparatively analyzed by HPLC. The HPLC analysis revealed one peak with a retention time (RT) of 5.5 min at an absorbance of 280 nm which was significantly increased for the *S. ambofaciens* pGM190/*papR2* sample compared to the pGM190 control (Figure 3.16, box). For compound purification purposes, a preparative HPLC was conducted and the fraction containing the peak at RT 5.5 min was sampled. The fraction was tested for bioactivity against *M. luteus,* where it showed a clear inhibition zone indicating that the obtained fraction harbored the bioactive metabolite whose production has been activated upon PapR2 induction.

AntiSMASH analysis with the *S. ambofaciens* genome sequence revealed 27 BGCs in total (Supplementary Figure S4). *S. ambofaciens* is a known producer of several different natural products, as for example spiramycin, congocidine, also called netropsin, stambomycin and others (Aigle et al., 2014), of which the coding BGCs correlate to regions 14, 20, and 24, respectively (Supplementary

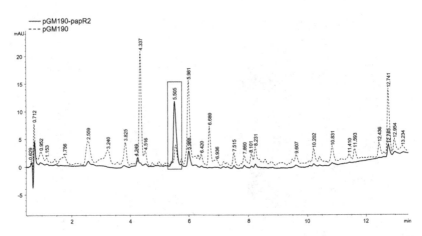

Figure 3.16 Comparative HPLC chromatogram of R5-derived ethyl acetate extracts of *S. ambofaciens* DSM40697 pGM190/*papR2* (solid line) and pGM190 (dotted line) separated with H$_2$O-MeOH gradient. Absorbance was measured at 280 nm. The box shows the fraction that was sampled

Figure S4). Two BGCs (region 2 and region 26; Supplementary Figures S5 and S6) were found to each contain three genes coding for SARPs, suggesting that these are the candidate BGCs activated by PapR2 expression (Fig. 9 A). The two BGCs show high overall similarity to each other and are located in the long terminal inverted repeats of the chromosome, which indicates a BGC duplication event. AntiSMASH analysis predicts that both BGCs are type II polyketide BGCs with 65 % similarity to a fluostatins M-Q BGC (Supplementary Figure S4).

Literature search revealed that a duplicated BGC, named *alp,* is present in *S. ambofaciens* and is responsible for the biosynthesis of the angucyclinone-type antibiotic kinamycin, which is a type II polyketide (Bunet et al., 2011). The BGC contains five regulatory genes, of which three, *alpT, alpU,* and *alpV,* encode SARP-family transcriptional regulators (Aigle et al., 2014). By manual gene cluster comparison of the SARP containing BGC region 2 identified by antiSMASH and the described kinamycin BGC, it was found that region 2 and 26 indeed constitute kinamycin BGCs, which are known to be responsible for the production of kinamycins C, D, and epoxy-kinamycin FL120B' (Aigle et al., 2014). Kinamycins were described to have antibiotic properties against *B. subtilis* but also being rapidly converted into an orange pigment without bioactivity (Bunet et al., 2011), which is consistent with the observed inconsistent bioactivity against gram-positive bacteria. A comparison of the amino acid sequences of PapR2 and the three SARP regulatory proteins AlpT, AlpU, and AlpV of the kinamycin BGC revealed that PapR2 is most similar to AlpU with 57% identity and 66% similarity scores and to a slightly lower extent similar to AlpT (58% similarity and 57 % identity) and AlpV (38% similarity and 48 % identity) (Supplementary Table S1). AlpV has been reported as the essential positive regulator for the activation of the *alp* cluster in *S. ambofaciens*, whereas no experimental data are available for AlpU and AlpT (Aigle et al., 2005). Thus, it can be assumed that PapR2 substituted for the functional role of one of the three SARP-type regulators of the *alp* BGC of *S. ambofaciens* and activated kinamycin biosynthesis.

3.4.7   *Streptomyces sp.* TÜ4106

To test the effect of *papR2* expression on secondary metabolite production, *Streptomyces* sp. TÜ4106 pGM190/*papR2* and the control strain *Streptomyces* sp. TÜ4106 pGM190 were cultivated in R5 medium as four independent biological replicates and in OM medium as three independent biological replicates. The time point of maximum bioactivity against test strains was determined at 48 h of cultivation. *Streptomyces* sp. TÜ4106 pGM190/*papR2* consistently caused a change of R5 medium coloration to blue while the *Streptomyces* sp. TÜ4106 pGM190 control retained the yellowish R5 medium color (Figure 3.17).

pGM190   pGM190/*papR2*

Figure 3.17 *Streptomyces* sp. TÜ4106 pGM190 (left) and pGM190/*papR2* (right) cultivated 72 h in R5 medium

Supernatant samples obtained from *Streptomyces* sp. TÜ4106 pGM190/*papR2* and pGM190 cultivated in R5 medium displayed bioactivity against *E. coli* mutant strains Δ*tolC* and Δ*lpxC*, and against *B. cinerea*, whereby there was no obvious difference between TÜ4106 pGM190/*papR2* and pGM190 samples (Supplementary Table S19). R5-derived supernatant samples and ethyl acetate extracts from TÜ4106 pGM190/*papR2* displayed increased bioactivity against *M. luteus* compared to the control TÜ4106 pGM190 (Figure 3.18 A, B). The average width of the inhibition halos clearly demonstrate a stronger inhibition of R5-derived supernatant samples of TÜ4106 pGM190/*papR2* with 10.74 ± 1.39 mm compared to the control TÜ4106 pGM190 with 4.10 ± 1 mm. A similar distribution is shown for R5-derived ethyl acetate extracts with an average width of 6.21 ± 2.9 mm for TÜ4106 pGM190/*papR2* and 1.67 ± 1.22 mm for TÜ4106 pGM190 (Figure 3.18C). Furthermore, the quantitative analyses show a relatively low variability in inhibition halo widths indicating that samples of TÜ4106 pGM190/*papR2* consistently exhibited stronger bioactivity against *M. luteus* compared to samples of TÜ4106 pGM190. R5-derived supernatant samples of TÜ4106 pGM190/*papR2* and pGM190 were tested against *B. subtilis* reporter strains, which did not indicate a specific mode of action for bioactive compounds (H. Brötz-Oesterhelt, pers. communication).

Antifungal activity was detected against *B. cinerea* from R5- and OM-derived supernatant samples and extracts, however, there was mostly no difference between TÜ4106 pGM190/*papR2* and TÜ4106 pGM190 samples (Supplementary Tables S19 and S20). The IBWF detected antifungal activity against *P. infestans*, which was displayed by ethyl acetate extract samples obtained from a TÜ4106 pGM190/*papR2* culture cultivated 48 h in R5 while the respective control extracts obtained of TÜ4106 pGM190 did not display any bioactivity (Supplementary Table S36). Additionally, general low activity against *P. infestans* was detected from supernatant samples of TÜ4106 pGM190/*papR2* and pGM190 (Supplementary Table S37).

To identify the secondary metabolite with bioactivity against *M. luteus*, the R5-derived supernatant and ethyl acetate extract samples of *Streptomyces* sp. TÜ4106 pGM190/*papR2* and as comparison from *Streptomyces* sp. TÜ4106 pGM190 (two replicates each) were analyzed by HPLC. The HPLC analysis of supernatant samples revealed multiple peaks at an absorbance of 280 nm which were significantly increased for TÜ4106 pGM190/*papR2* (Figure 3.19A) compared to the TÜ4106 pGM190 control (Figure 3.19B). To identify the bioactive compound, a preparative HPLC was conducted and the sample from 0.5–12.5 min was separated every minute into distinct fractions. All fractions obtained from supernatant and extract samples of TÜ4106 pGM190/*papR2* and pGM190 were tested against *M. luteus*.

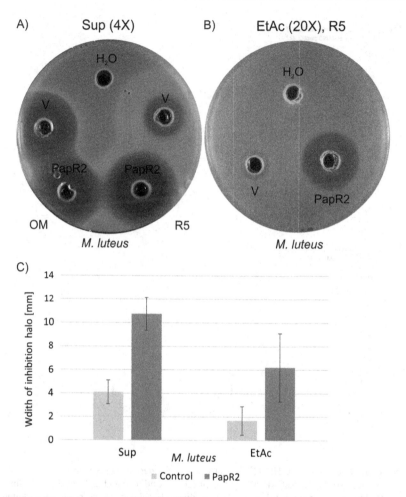

Figure 3.18   Bioassay results of TÜ4106 control strain (pGM190, "V") and *papR2* overexpression strain (pGM190/*papR2*, "PapR2") after 48 h cultivation in R5 and OM medium. **A)** R5- and OM-derived 4-fold concentrated supernatant ("sup") was tested against *M. luteus*. **B)** R5-derived 20-fold concentrated ethyl acetate extract ("EtAc") in 50 % MeOH was tested against *M. luteus*. **C)** Average width of inhibition halos of R5-derived supernatant (4X, N = 3) and extract (20X, N = 3) against *M. luteus*. Analysis was conducted with ImageJ and black lines show the standard deviation. N = Number of replicates

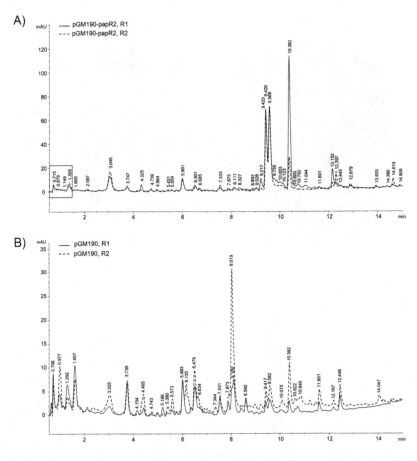

Figure 3.19 Comparative HPLC chromatogram of R5-derived supernatant samples of **A)** *Streptomyces* sp. TÜ4106 pGM190/*papR2* replicate 1 (solid line) and replicate 2 (dotted line) and **B)** *Streptomyces* sp. TÜ4106 pGM190 replicate 1 (solid line) and replicate 2 (dotted line) separated with H_2O-MeOH gradient. Absorbance was measured at 280 nm. The box indicates the fraction containing the bioactive compound

The first fraction of the supernatant of TÜ4106 pGM190/*papR2* (Figure 3.19A, box) showed inhibition of *M. luteus* (Figure 3.20A). Simultaneously, the first fraction of the ethyl acetate extract sample of TÜ4106 pGM190/*papR2* and the first fraction of the supernatant sample of the TÜ4106 pGM190 control did not show any bioactivity against *M. luteus* (Figure 3.20B). This implies that the bioactive compound is inorganic.

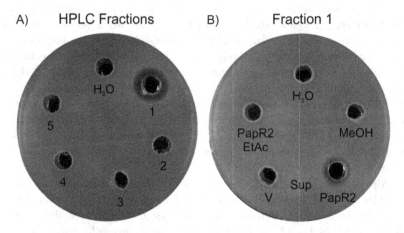

Figure 3.20   Bioassay results of TÜ4106 obtained fractions from R5-derived supernatants and extracts tested against *M. luteus* with H_2O and 50 % MeOH as negative controls. **A)** 7-fold concentrated supernatant sample ("sup") from culture grown in R5 medium fractions 1–5 (0.5–5.5 min). **B)** Fraction 1 from different samples: supernatant of TÜ4106 pGM190/ *papR2* ("Sup", "PapR2", same sample as Fraction 1 in A), supernatant of TÜ4106 pGM190 ("Sup", "V"), extract from TÜ4106 pGM190/*papR2* ("PapR2 EtAc", same culture as "Sup PapR2")

AntiSMASH analysis with the *Streptomyces* sp. TÜ4106 genome sequence revealed 17 BGCs in total (Supplementary Figure S39). Two BGCs were found to contain four genes coding for SARPs of which three are located in BGC region 1.3 and one is located within the BGC region 1.17 (Supplementary Figures S40 and S41). A BLASTP comparison revealed that the amino acid sequence of the putative SARP gene *moaR1* within region 1.17 displays with 58 % the highest similarity to the PapR2 amino acid sequence, suggesting a high probability to be activated by PapR2 (Supplementary Table S7). The respective BGC was predicted to belong to several BGC types including a type-II PKS. TÜ4106 pGM190/*papR2* produced a blue pigment during cultivation, thus, the production of a type-II polyketide is likely because type II polyketides are often optically active due to their basic aromatic structure. This implies that BGC 1.17 might be activated by PapR2 and the observed pigmented, bioactive compound might be the product of a type-II PKS.

3.5   The modified PapR2 protein causes a shift of the *papR1* promoter

Besides the potential of PapR2 to activate silent BGCs, mechanistic studies on the SARP-type regulators shall be carried out in this work. So far, very little is known about the molecular principle of SARP-type regulators. Furthermore, the crystal structure of the SARP-type is not elucidated yet and nothing is known regarding the protein conformation status or potential ligands that bind to SARPs. The most similar analyzed crystal structure of a SARP-like protein is that of EmbR from *Mycobacterium tuberculosis* (Alderwick et al., 2006). To get a better understanding of the protein structure of SARP-type regulators, bioinformatics analysis with SARP amino acid sequences and the protein structure prediction tool AlphaFold was conducted. Furthermore, the tool I-Tasser was used to be estimate potential ligands of PapR2.

AlphaFold analysis revealed that the bacterial transcriptional activation domain (BTAD) and DNA binding domain (DBD) of PapR2 consist of α-helical structures (Figure 3.21A). Comparisons with the crystal structure of EmbR suggest that the DBD of PapR2 might consist of a central three-helix bundle flanked by two β-sheets similar to EmbR (Alderwick et al., 2006; Wietzorrek & Bibb, 1997). On the contrary, the BTAD of EmbR consists of eight α-helices (Alderwick et al., 2006), while the AlphaFold analysis proposed only three α-helices for the BTAD of PapR2 (Figure 3.21A). I-Tasser predicted "nucleic acid" and "peptides" as potential ligands for PapR2 (data not shown). The potential ligands were predicted to bind to similar regions within the putative BTAD of PapR2. Interestingly, binding of peptidyl ligands was shown for the SARP NosP in *Streptomyces actuosus* (Li et al., 2018). Thus, it might be possible that PapR2 may also bind peptidyl ligands as effectors.

Furthermore, AlphaFold analysis predicted an N-terminal extension with a final α-helix for PapR2 (Figure 3.21B). It was hypothesized that this structural region is not necessary for the function of PapR2 and may rather disturb crystallization attempts. Hence, a shortened version of PapR2, as well as the complete PapR2 were generated to crystallize PapR2 and experimentally confirm the protein tertiary structure. The native *papR2* sequence, as well as the sequenced shortened by 257 bp have been synthesized as *E. coli* codon usage optimized sequences and cloned into pET28(+) vectors, resulting in the constructs pET28(+)_papR2 and pET28(+)_papR2_257, respectively (GenScript Biotech). The proteins were coupled to a Strep-Tag resulting in a complete Strep-tagged PapR2 protein consisting of 356 amino acids and a 257 bp-shortened Strep-tagged PapR2 derivative consisting of 282 amino acids and lacking the N-terminal extension (Figure 3.21B). The constructs were used for *papR2* and *papR2–257* expression in *E. coli* BL21. The design of the *papR2* pET28-constructs, expression in *E. coli* and protein purification was performed by Konrad Büssow (AG Wulf Blankenfeldt, HZI) and the two purified PapR2 derivatives were kindly provided for DNA binding studies.

A)   PapR2 monomer   B)   N-terminal extension

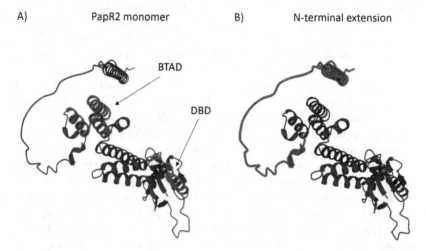

Figure 3.21   Structure predictions of **A)** the native PapR2 as monomer with functional domains (bacterial transcriptional activator domain (BTAD) and DNA binding domain (DBD)), **B)** PapR2 monomer with N-terminal extension, which was deleted in the 257 bp-shortened PapR2 derivative. Protein structure predictions were performed with AlphaFold and visualized with RCSB protein data bank Mol* 3D Viewer

To test the functionality of the *E. coli*-derived PapR2 proteins, the DNA binding capacity of two replicates of the complete Strep-tagged PapR2 ("R1" and "R2", size: ~38.6 kDa) and Strep-tagged 257 bp-shortened PapR2 with the incomplete amino acid sequence ("D-257", size: ~31.0 kDa) was assessed by electrophoretic mobility shift assays (EMSA) in 2 % agarose gels. In previous studies it was found that PapR2, which was obtained as His-tagged protein via heterologous expression in *S. lividans*, binds to the *papR1* promoter of *S. pristinaespiralis* (Mast et al., 2015) as well as to the RedD-controlled promoters *redP* and *redQ* of *S. lividans* (Krause et al., 2020). R1, R2, and D-257 were tested for binding to pro*papR1*, which was labelled with Cy5 for visualization (Figure 3.22A). Only the second replicate of the complete Strep-tagged PapR2

("R2") caused a shift of pro*papR1* indicating a potential protein-DNA interaction, whereas a shift was not visible for the first replicate of complete Strep-tagged PapR2 ("R1") and the 257 bp-shortened PapR2 derivative ("D-257"). The SDS PAGE revealed that the "R2" sample causing the promoter shift is the only sample containing complete Strep-tagged PapR2 protein at the expected size of ~38.6 kDa (Figure 3.22B). SDS PAGE of R1 and D-257 did not show visible protein bands at the expected sizes of ~38.6 kDa and ~31 kDa, respectively. Thus, the Strep-tagged *E. coli*-derived PapR2 protein with the complete amino acid sequence is probably able to bind the *papR1* promoter. It was not possible to evaluate the binding capacity of the 257 bp-shortened PapR2 derivative (D-257) since protein abundance could not be confirmed with SDS PAGE analysis.

To confirm that the second replicate of Strep-tagged complete PapR2 (R2) specifically binds to the *papR1* promoter, competitive EMSAs were performed with Cy5-labeled pro*papR1*, the PapR2 protein (R2) and increasing concentrations of competitive, unlabeled pro*papR1*. If protein-DNA interaction is specific, increasing concentrations of competitive, unlabeled pro*papR1* will increasingly bind PapR2 resulting in the gel retardation of Cy5-pro*papR1* being brought back to the running length of the "reference-DNA" ("C", Cy5-pro*papR1* alone). EMSA results indicated that the gel retardation of Cy5-pro*papR1* was partly brought back to the running length of the reference DNA upon application of competitive, unlabeled pro*papR1* (Figure 3.22C). An PCR amplificate of the *aphII* gene served as negative control as 2 μL of the DNA was applied to a sample containing Cy5-pro*papR1* and PapR2 protein (R2). A shift of pro*papR1* was expected similar to the "R2" lane (Figure 3.22C), however, the gel retardation of Cy5-pro*papR1* was completely brought back to the running length of the reference ("C") as it was expected for the samples containing competitive, unlabeled pro*papR1*. It remains unclear why the sample with the unspecific DNA of the *aphII* gene, which does not contain the binding motif sequence of PapR2, did not show a promoter shift of *papR1*. A possible explanation might be that the PCR amplificate disturbs interaction of PapR2 and pro*papR1*. Furthermore, the bands of pro*papR1* potentially bound by PapR2 were also not clearly visible as only the reference DNA was always visible. Thus, EMSA analyses should be repeated with a native polyacrylamide gel instead of an 2 % agarose gel as it might present clearer results. Furthermore, newly expressed proteins of complete Strep-tagged PapR2 and 257 bp-shortened Strep-tagged PapR2 derivative should be obtained to confirm the specificity of complete Strep-tagged PapR2 interaction with pro*papR1* and to analyze if the 257 bp-shortened Strep-tagged PapR2 derivative is also able to bind pro*papR1*. Confirming the DNA binding capacity of the *E. coli*-derived PapR2 proteins will allow facilitated production of the PapR2 proteins, which will enable studies on protein structure and molecular mechanism and help for a better understanding of the function of SARPs.

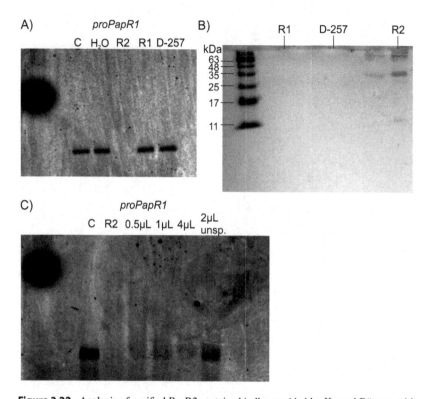

Figure 3.22 Analysis of purified PapR2 proteins kindly provided by Konrad Büssow with the following samples: R1 & R2 = Strep-tagged complete PapR2 replicates 1 and 2; D-257 = 257 bp-shortened Strep-tagged PapR2 derivative; C = control, Cy5-*papR1* promoter (4 ng/μL). **A)** EMSA of received protein samples: 1 μL Cy5-pro*papR1*, 5 μL H$_2$O or protein. **B)** SDS PAGE of received protein samples. **C)** Competitive EMSA of 5 μL Strep-tagged PapR2 replicate 2 (R2), 1 μL Cy5-pro*papR1* and increasing volumes (0.5–4 μL) of unlabeled pro*papR1* (122 ng/μL). "2 μL unsp." = unspecific PCR amplificate of the *aphII* gene as control (50 ng/μL)

Discussion

4

In this study, SARP-driven expression of BGCs of a selected number of genome-sequenced actinomycetes strains was successfully achieved. Overexpression of PapR2 resulted in increased production of bioactive compounds which were assessed with bioassays against different test strains including bacteria and fungi. Thus, it was verified that the targeted SARP-driven activation strategy is a valuable method to be applied to unlock the genetic potential of actinomycetes to produce novel natural products. Development of computational tools has been a key technology for the search of potentially novel natural products by mining the abundant microbial genomic sequence information (Locey & Lennon, 2016; Nayfach et al., 2021). Several software tools were established to analyze genomes, find potential BGCs and even predict the chemistry of the metabolic product (Hemmerling & Piel, 2022). These bioinformatics tools have helped to reveal the great genetic potential of actinomycetes to synthesize far more than the metabolites readily produced under standard laboratory cultivation conditions (Gavriilidou et al., 2022).

This study aimed at using overexpression of the pathway-specific transcriptional activator PapR2 (from the pristinamycin BGC in *S. pristinaespiralis*) to induce the expression of silent BGCs in other species of the *Streptomycetaceae* family, by PapR2 taking over the activating function of the native SARP(s). This requires PapR2 to recognize the DNA sequence to which the native SARP binds. Therefore, and as it was suggested in the previous study of Krause et al. (2020), it is more likely to succeed when the target strain carries a BGC regulated by a SARP with high similarity to PapR2 and whose binding motif sequence is also highly similar to the known binding motif of PapR2. The target strains in this study were prioritized according to the criteria of availability in DSMZ collection, occurrence of PapR2 homologs, and the PapR2 binding motifs within the genome. antiSMASH 6.0 (Blin et al., 2021) was used to identify putative BGCs encoded within the genome and to predict which BGCs might be activated upon overexpression of *papR2*. It was expected that the BGC containing genes which encode SARPs similar to PapR2 were most likely the ones to be affected, which provides a starting point to associate BGCs with an observed compound or activity.

The approach to activate secondary metabolism by overexpression of transcriptional activators has been established in modern times as genome mining became popular during the search for novel chemical compounds. In some studies, activation or increased gene expression of BGCs was achieved by overexpression of native transcriptional activators. For instance, overexpression of a MarR-family transcriptional regulator increased avermectin production in *S. avermitilis* (Guo et al., 2018), overexpression of the TetR-like regulator gene *sabR* increased nikkomycin biosynthesis in *S. ansochromogenes* (Pan et al., 2011), and constitutive expression of a gene encoding a LAL protein found in a silent, unknown BGC in *S. ambofaciens* resulted in discovery of stambomycins A-D (Laureti et al., 2011). In the current study, the focus was on the transcriptional activators of the SARP family which are widely distributed among actinomycetes with especially high abundance among streptomycetes. Activation of BGCs by overexpression of a panel of native SARPs was reported for *S. rochei* (Misaki et al., 2022) and for *Streptomyces* sp. MSC090213JE08 coupled with an OSMAC approach (Du et al., 2016), which resulted in identification of a novel cyclohexene-containing enamide and a novel amide-containing polyene, respectively. Transcriptional activation of BGCs by overexpression of native SARPs has the advantage of a higher SARP specificity which might promise higher success rates. However, only focusing on native SARPs at each strain would slow down the discovery rate of new secondary metabolites. Since SARPs share relatively similar protein architectures and bind similar recognition motifs, the possibility for SARPs substituting other SARPs in heterologous hosts is relatively high. This has been shown for PapR2 taking over the regulatory function of RedD, the native SARP in *S. lividans* (Krause et al., 2020). In the current study, it was confirmed that PapR2 can substitute the regulatory function of RedD also in *S. coelicolor* A3(2). Furthermore, other studies also verified that heterologous expression of SARPs can affect secondary metabolism by complementing the function of native SARPs (Beck et al., 2021; Garg & Parry, 2010; Krause et al., 2020). Thus, selection of a candidate SARP for heterologous expression can be applied to a number of actinomycetes species and not just one native host, which could significantly accelerate the discovery of new secondary metabolites. This was confirmed in the current study as heterologous expression of the SARP transcriptional activator PapR2 positively affected biosynthesis of secondary metabolites with bioactivity against different microbial test strains in more than half of selected actinomycetes strains. PapR2 might have specifically activated the kinamycin BGC in *S. ambofaciens* DSM 40697, a hypothesis that is currently being tested by LC-MS analysis of the bioactive fraction. Furthermore, heterologous expression of PapR2 in the strain *Streptomyces griseus* S4-7 also helped to activate the BGC encoding for a novel post-translationally modified peptide (RiPPs) S4-7 with anti-*Fusarium* activity (Soto Zarazua, pers. comm.; AG Max Crüsemann from the Institute for Pharmaceutical Biology at University of Bonn). This relative high success rate supports the aforementioned strategy for prioritization of strains based on bioinformatics analysis of genome sequences.

It still needs to be elucidated which key features enable substitution of SARPs. In this study, it was shown that PapR2 is able to substitute RedD in *S. coelicolor* A3(2) but not ActII-ORF4 and AfsR. One key feature for complementation might be associated with the group of SARPs. PapR2, RedD and Act-ORF4 belong to the "small" SARPs while AfsR belongs to the "large" SARP group, which might explain why PapR2 cannot substitute for AfsR. This hypothesis is supported by another study which demonstrated that AfsR cannot substitute for either RedD or ActII-ORF4 in the respective *S. coelicolor* deletion mutants (Floriano & Bibb, 1996). However, belonging to the same group of SARPs is not sufficient as PapR2 was not able to complement ActII-ORF4 in *S. coelicolor* A3(2). Further key features might be homology of nucleotide or amino acids sequences, sufficiently similar binding motifs, or homology of protein structures. The current study used the similarity of amino acid sequences of SARPs as indicator for possible complementation, however, this has not been verified as critical requirement. Garg & Parry (2010) demonstrated that the SARP transcriptional activator VlmI from *S. viridifaciens* can substitute RedD but not ActII-ORF4 the *S. coelicolor* M512 mutant strain. Interestingly, a BLASTP comparison of VlmI with RedD and ActII-ORF4 amino acid sequences revealed very close similarity scores of 49% and 48%, respectively (Accessions: AAN10246.1 (VlmI) and Table 3.1). Here, the authors stated that complementation of RedD might be enabled due to both proteins containing a similarly extended N-terminus which is not present ActII-ORF4 (Garg & Parry, 2010). PapR2 has been proven to be a good candidate for SARP-driven activation of BGCs by heterologous expression in different actinomycetes. However, higher success rates might be achieved and target BGCs accurately determined when SARPs, especially PapR2, are further mechanistically studied.

Although numerous SARP-type regulators are known and their positive effect on transcription of industrially and clinically important substances has been observed, there have been only a few comprehensive mechanistic studies of

SARP regulators performed to date. Investigating the biological mechanisms of SARP-driven regulation would help to better identify which silent BGCs might be activated and to select candidate strains. Some of the questions about the molecular mechanisms of SARP regulators that still need to be elucidated are 1) which amino acids are essential for transcriptional activation, 2) which nucleotides of the binding motif are essential for SARP binding, 3) what is the protein tertiary structure, and 4) what is the oligomerization state of the active regulator.

Crystallization of OmpR provided a first model for the structure of SARPs. The crystal structure of OmpR revealed a C-terminal winged helical structure which was identified as DNA binding domain (DBD) (Martínez-Hackert & Stock, 1997). The DBD of OmpR was the key homology feature found also to be conserved in SARPs at the N-terminus and used to originally propose the SARP family of regulators (Wietzorrek & Bibb, 1997). Thus, it is likely that the overall tertiary structure of OmpR DBD is shared by SARPs. In more recent studies, the molecular structure of EmbR, a transcriptional activator of *Mycobacterium tuberculosis* related to the SARP family of transcriptional factors, was published (Alderwick et al., 2006). The crystallization studies of EmbR indeed revealed an N-terminal DBD similar to OmpR as predicted for SARPs and an adjacent bacterial transcriptional activation domain (BTAD) (Alderwick et al., 2006). Based on the known protein structures and alignments with amino acid sequences of SARP proteins, SARPs were suggested to contain a N-terminal OmpR-like DBD, an adjacent BTAD domain and a C-terminal domain which may contain additional domains (Alderwick et al., 2006; Liu et al., 2013; Wietzorrek & Bibb, 1997). However, experimental confirmation of the protein structure of SARPs is still necessary. Studies of AfsR, a SARP transcriptional regulator of *S. coelicolor* A3(2), provided a first model for transcriptional activation by SARPs (Tanaka & Omura, 1990). Sequence alignment of AfsR target promoters revealed the presence of direct heptameric repeats indicating a cooperative binding of two SARPs, which was experimentally confirmed for AfsR (Tanaka & Omura, 1990). Since these heptameric repeats were found in multiple promoter sequences of SARP target genes, the formation of dimers for DNA binding was suggested as general transcriptional activation mechanisms among the SARP family. This manner is also aimed to be investigated for PapR2 by Bacterial-Two-Hybrid assays, which indicate interaction between two proteins. Preparative work regarding cloning procedures for the Bacterial-Two-Hybrid assays already started in this study (data not shown) and will be the basis to further uncover the molecular mechanism of PapR2. Furthermore, a study on DNA binding capacity of the DnrI SARP regulator of *S. peucetius* provided evidence for critical amino acids within the

N-terminal DBD and critical nucleotides within the DNA tandem repeats (Sheldon et al., 2002). All these mechanistic studies on regulators belonging or related to the SARP family of transcriptional activators provide models for molecular mechanisms of how transcriptional activation is achieved. However, comprehensive mechanistic studies on SARPs are still pending. Nevertheless, mechanistic studies on PapR2 as a model for SARPs are in the pipeline and PapR2 derivatives modified for *E. coli* expression to facilitate protein production have already been designed. This study showed that the modified PapR2 derivative is still functional in its DNA binding capacity and therefore might be suitable as model to investigate the tertiary structure of SARP regulators.

References

Aigle, B., Lautru, S., Spiteller, D., Dickschat, J. S., Challis, G. L., Leblond, P. & Pernodet, J.-L. (2014). Genome mining of *Streptomyces ambofaciens*. *J. Ind. Microbiol. Biotechnol.*, *41*(2), 251–263. https://doi.org/10.1007/s10295-013-1379-y

Aigle, B., Pang, X., Decaris, B. & Leblond, P. (2005). Involvement of AlpV, a new member of the *Streptomyces* antibiotic regulatory protein family, in regulation of the duplicated type II polyketide synthase alp gene cluster in *Streptomyces ambofaciens*. *J. Bacteriol.*, *187*(7), 2491–2500. https://doi.org/10.1128/JB.187.7.2491-2500.2005

Alderwick, L. J., Molle, V., Kremer, L., Cozzone, A. J., Dafforn, T. R., Besra, G. S. & Fütterer, K. (2006). Molecular structure of EmbR, a response element of Ser/Thr kinase signaling in *Mycobacterium tuberculosis*. *PNAS*, *103*(8), 2558–2563. https://doi.org/10.1073/pnas.0507766103

Arias, P., Fernández-Moreno, M. A. & Malpartida, F. (1999). Characterization of the pathway-specific positive transcriptional regulator for actinorhodin biosynthesis in *Streptomyces coelicolor* A3(2) as a DNA-binding protein. *J. Bacteriol.*, *181*(22), 6958–6968. https://doi.org/10.1128/JB.181.22.6958-6968.1999

Baltz, R. H. (2017). Gifted microbes for genome mining and natural product discovery. *J. Ind. Microbiol. Biotechnol.*, *44*(4–5), 573–588. https://doi.org/10.1007/s10295-016-1815-x

Barka, E. A., Vatsa, P., Sanchez, L., Gaveau-Vaillant, N., Jacquard, C., Meier-Kolthoff, J. P., Klenk, H.-P., Clément, C., Ouhdouch, Y. & van Wezel, G. P. (2016). Taxonomy, Physiology, and Natural Products of Actinobacteria. *Microbiol. Mol. Bio. Rev.*, *80*(1), 1–43. https://doi.org/10.1128/MMBR.00019-15

Baron, S. S. & Rowe, J. J. (1981). Antibiotic action of pyocyanin. *Antimicrob. Agents Chemother.*, *20*(6), 814–820. https://doi.org/10.1128/aac.20.6.814

Bate, N., Butler, A. R., Gandecha, A. R. & Cundliffe, E. (1999). Multiple regulatory genes in the tylosin biosynthetic cluster of *Streptomyces fradiae*. *Chem. Biol.*, *6*(9), 617–624. https://doi.org/10.1016/s1074-5521(99)80113-6

Beck, C., Gren, T., Ortiz-López, F. J., Jørgensen, T. S., Carretero-Molina, D., Martín Serrano, J., Tormo, J. R., Oves-Costales, D., Kontou, E. E., Mohite, O. S., Mingyar, E., Stegmann, E., Genilloud, O. & Weber, T. (2021). Activation and Identification of a Griseusin Cluster in *Streptomyces* sp. CA-256286 by Employing Transcriptional Regulators and Multi-Omics Methods. *Molecules*, *26*(21). https://doi.org/10.3390/molecules26216580

© The Editor(s) (if applicable) and The Author(s), under exclusive license to Springer Fachmedien Wiesbaden GmbH, part of Springer Nature 2024
O. Rössler, *SARP-Driven Activation of Antibiotic Biosynthetic Gene Clusters in Actinomycetes*, BestMasters, https://doi.org/10.1007/978-3-658-44552-2

Bentley, S. D., Chater, K. F, Cerdeño-Tárraga, A.-M., Challis, G. L, Thomson, N. R., James, K. D., Harris, D. E., Quail, M. A., Kieser, H., Harper, D., Bateman, A., Brown, S., Chandra, G., Chen, C. W., Collins, M., Cronin, A., Fraser, A., Goble, A., Hidalgo, J., . . . Hopwood, D. A. (2002). Complete genome sequence of the model actinomycete *Streptomyces coelicolor* A3(2). *Nature*, *417*(6885), 141–147. https://doi.org/10.1038/417141a

Bérdy, J. (2012). Thoughts and facts about antibiotics: where we are now and where we are heading. *J. Antibiot.*, *65*(8), 385–395. https://doi.org/10.1038/ja.2012.27

Bertasso, M., Holzenkämpfer, M., Zeeck, A., Dall'Antonia, F. & Fiedler, H. P. (2001). Bagremycin A and B, novel antibiotics from *streptomyces* sp. Tü 4128. *J. Antibiot.*, *54*(9), 730–736. https://doi.org/10.7164/antibiotics.54.730

Bibb, M. J. (2005). Regulation of secondary metabolism in streptomycetes. *Curr. Opin. Microbiol.*, *8*(2), 208–215. https://doi.org/10.1016/j.mib.2005.02.016

Bibb, M. J., Ward, J. M. & Hopwood, D. A. (1978). Transformation of plasmid DNA into *Streptomyces* at high frequency. *Nature*, *274*(5669), 398–400. https://doi.org/10.1038/274 398a0

Bibb, M. J., White, J., Ward, J. M. & Janssen, G. R. (1994). The mRNA for the 23S rRNA methylase encoded by the ermE gene of *Saccharopolyspora erythraea* is translated in the absence of a conventional ribosome-binding site. *Mol. Microbiol.*, *14*(3), 533–545. https://doi.org/10.1111/j.1365-2958.1994.tb02187.x.

Blin, K., Shaw, S., Kloosterman, A. M., Charlop-Powers, Z., van Wezel, G. P., Medema, M. H. & Weber, T. (2021). antiSMASH 6.0: improving cluster detection and comparison capabilities. *Nucleic Acids Res.*, *49*(W1), W29-W35. https://doi.org/10. 1093/nar/gkab335

Bruyn, A. de, Verellen, S., Bruckers, L., Geebelen, L., Callebaut, I., Pauw, I. de, Stessel, B. & Dubois, J. (2022). Secondary infection in COVID-19 critically ill patients: a retrospective single-center evaluation. *BMC Infect. Dis.*, *22*(1), 207. https://doi.org/10.1186/s12879-022-07192-x

Bunet, R., Song, L., Mendes, M. V., Corre, C., Hotel, L., Rouhier, N., Framboisier, X., Leblond, P., Challis, G. L. & Aigle, B. (2011). Characterization and manipulation of the pathway-specific late regulator AlpW reveals *Streptomyces ambofaciens* as a new producer of Kinamycins. *J. Bacteriol.*, *193*(5), 1142–1153. https://doi.org/10.1128/JB.012 69-10

Carlson, C. J., Albery, G. F., Merow, C., Trisos, C. H., Zipfel, C. M., Eskew, E. A., Olival, K. J., Ross, N. & Bansal, S. (2022). Climate change increases cross-species viral transmission risk. *Nature*, *607*(7919), 555–562. https://doi.org/10.1038/s41586-022-047 88-w

Castro, J. F., Razmilic, V., Gomez-Escribano, J. P., Andrews, B., Asenjo, J. A. & Bibb, M. J. (2015). Identification and Heterologous Expression of the Chaxamycin Biosynthesis Gene Cluster from *Streptomyces leeuwenhoekii*. *Appl. Environ. Microbiol.*, *81*(17), 5820–5831. https://doi.org/10.1128/AEM.01039-15

CDC Centers for Disease Control and Prevention (U.S.), National Center for Emerging Zoonotic and Infectious Diseases & Division of Healthcare Quality Promotion. Antibiotic Resistance Coordination and Strategy Unit. (2019). *Antibiotic resistance threats in the United States, 2019*. https://doi.org/10.15620/cdc:82532

Chen, I.-M. A., Chu, K., Palaniappan, K., Pillay, M., Ratner, A., Huang, J., Huntemann, M., Varghese, N., White, J. R., Seshadri, R., Smirnova, T., Kirton, E., Jungbluth, S. P.,

Woyke, T., Eloe-Fadrosh, E. A., Ivanova, N. N. & Kyrpides, N. C. (2019). IMG/M v.5.0: an integrated data management and comparative analysis system for microbial genomes and microbiomes. *Nucleic Acids Res.*, *47*(D1), D666-D677. https://doi.org/10.1093/nar/gky901

Cohen, S. N., Chang, A. C. & Hsu, L. (1972). Nonchromosomal antibiotic resistance in bacteria: genetic transformation of *Escherichia coli* by R-factor DNA. *PNAS*, *69*(8), 2110–2114. https://doi.org/10.1073/pnas.69.8.2110

Covington, B. C., Xu, F. & Seyedsayamdost, M. R. (2021). A Natural Product Chemist's Guide to Unlocking Silent Biosynthetic Gene Clusters. *Annu. Rev. Biochem.*, *90*, 763–788. https://doi.org/10.1146/annurev-biochem-081420-102432

Cundliffe, E. (2006). Antibiotic production by actinomycetes: the Janus faces of regulation. *J. Ind. Microbiol. Biotechnol.*, *33*(7), 500–506. https://doi.org/10.1007/s10295-006-0083-6

Delcour, A. H. (2009). Outer membrane permeability and antibiotic resistance. *BBA*, *1794*(5), 808–816. https://doi.org/10.1016/j.bbapap.2008.11.005

Diana, P. & Cirrincione, G. Four-Membered Heterocyclic Rings and Their Fused Derivatives. In *Biosynthesis of Heterocycles* (S. 277–378). https://doi.org/10.1002/9781118960554.ch4

Diana, P. & Cirrincione, G. (Hrsg.). (2015). *Biosynthesis of Heterocycles*. John Wiley & Sons, Inc. https://doi.org/10.1002/9781118960554

Du, D., Katsuyama, Y., Onaka, H., Fujie, M., Satoh, N., Shin-Ya, K. & Ohnishi, Y. (2016). Production of a Novel Amide-Containing Polyene by Activating a Cryptic Biosynthetic Gene Cluster in *Streptomyces* sp. MSC090213JE08. *Chembiochem*, *17*(15), 1464–1471. https://doi.org/10.1002/cbic.201600167

Etebu, E. & Arikekpar, I. (2016). Antibiotics: Classification and mechanisms of action with emphasis on molecular perspectives. *IJAMBR*(4), 90–101.

Flett, F., Mersinias, V. & Smith, C. P. (1997). High efficiency intergeneric conjugal transfer of plasmid DNA from *Escherichia coli* to methyl DNA-restricting streptomycetes. *FEMS Microbiol. Lett.*, *155*(2), 223–229. https://doi.org/10.1111/j.1574-6968.1997.tb13882.x

Floriano, B. & Bibb, M. (1996). *afsR* is a pleiotropic but conditionally required regulatory gene for antibiotic production in *Streptomyces coelicolor* A3(2). *Mol. Microbiol.*, *21*(2), 385–396. https://doi.org/10.1046/j.1365-2958.1996.6491364.x

Garg, R. P. & Parry, R. J. (2010). Regulation of valanimycin biosynthesis in *Streptomyces viridifaciens*: characterization of VlmI as a *Streptomyces* antibiotic regulatory protein (SARP). *Microbiology (Reading, Engl.)*, *156*(Pt 2), 472–483. https://doi.org/10.1099/mic.0.033167-0

Gavriilidou, A., Kautsar, S. A., Zaburannyi, N., Krug, D., Müller, R., Medema, M. H. & Ziemert, N. (2022). Compendium of specialized metabolite biosynthetic diversity encoded in bacterial genomes. *Nat. Microbiol.*, *7*(5), 726–735. https://doi.org/10.1038/s41564-022-01110-2

Gomez-Escribano, J. P., Castro, J. F., Razmilic, V., Jarmusch, S. A., Saalbach, G., Ebel, R., Jaspars, M., Andrews, B., Asenjo, J. A. & Bibb, M. J. (2019). Heterologous Expression of a Cryptic Gene Cluster from *Streptomyces leeuwenhoekii* C34T Yields a Novel Lasso Peptide, Leepeptin. *Appl. Environ. Microbiol.*, *85*(23). https://doi.org/10.1128/AEM.01752-19

Gomez-Escribano, J. P., Song, L., Fox, D. J., Yeo, V., Bibb, M. J. & Challis, G. L. (2012). Structure and biosynthesis of the unusual polyketide alkaloid coelimycin P1, a metabolic product of the *cpk* gene cluster of *Streptomyces coelicolor* M145. *Chem. Sci., 3*(9), 2716. https://doi.org/10.1039/C2SC20410J

Guo, J., Zhang, X., Lu, X., Liu, W., Chen, Z., Li, J., Deng, L. & Wen, Y. (2018). SAV4189, a MarR-Family Regulator in *Streptomyces avermitilis*, Activates Avermectin Biosynthesis. *Front. Microbiol., 9*, 1358. https://doi.org/10.3389/fmicb.2018.01358

Hemmerling, F. & Piel, J. (2022). Strategies to access biosynthetic novelty in bacterial genomes for drug discovery. *Nat. Rev. Drug Discov., 21*(5), 359–378. https://doi.org/10.1038/s41573-022-00414-6

Hopwood, D. A., Bibb, M. J., Chater, K. F., Kieser, T., Bruton, C. J., Kieser, H. M., Lydiate, D. J., Smith, C. P., Ward, J. M. & Schrempf, H. (1985). Genetic manipulation of *streptomyces* — A laboratory manual. *Biochem. Educ.*(14), Artikel 4, 196. https://doi.org/10.1016/0307-4412(86)90228-1

Hopwood, D. A., Chater, K. F. & Bibb, M. J. (1995). Genetics of antibiotic production in *Streptomyces coelicolor* A3(2), a model streptomycete. *Biotechnol., 28*, 65–102. https://doi.org/10.1016/b978-0-7506-9095-9.50009-5

Hutchings, M. I., Truman, A. W. & Wilkinson, B. (2019). Antibiotics: past, present and future. *Curr. Opin. Microbiol., 51*, 72–80. https://doi.org/10.1016/j.mib.2019.-10.008

Igarashi, M., Kinoshita, N., Ikeda, T., Kameda, M., Hamada, M. & Takeuchi, T. (1997). Resormycin, a novel herbicidal and antifungal antibiotic produced by a strain of *Streptomyces platensis*. I. Taxonomy, production, isolation and biological properties. *J. Antibiot., 50*(12), 1020–1025. https://doi.org/10.7164/antibiotics.50.1020

Ikeda, H., Ishikawa, J., Hanamoto, A., Shinose, M., Kikuchi, H., Shiba, T., Sakaki, Y., Hattori, M. & Omura, S. (2003). Complete genome sequence and comparative analysis of the industrial microorganism *Streptomyces avermitilis*. *Nat. Biotechnol., 21*(5), 526–531. https://doi.org/10.1038/nbt820

Ju, J., Lim, S.-K., Jiang, H., Seo, J.-W. & Shen, B. (2005). Iso-migrastatin congeners from *Streptomyces platensis* and generation of a glutarimide polyketide library featuring the dorrigocin, lactimidomycin, migrastatin, and NK30424 scaffolds. *J. Am. Chem. Soc., 127*(34), 11930–11931. https://doi.org/10.1021/ja053118u

Katz, L. & Baltz, R. H. (2016). Natural product discovery: past, present, and future. *J. Ind. Microbiol. Biotechnol., 43*(2–3), 155–176. https://doi.org/10.1007/s10295-015-1723-5

Kieser, T., Bibb, M. J., Chater, K. F., Buttner, M. J. & Hopwood, D. A. (2000). *Practical streptomyces genetics*. John Innes Foundation.

Krause, J., doctoral dissertation (2021). Isolierung neuer Antibiotika-Produzenten aus indonesischen Bodenproben und Aktivierung stiller Antibiotika-Gencluster. https://doi.org/10.15496/publikation-36593

Krause, J., Handayani, I., Blin, K., Kulik, A. & Mast, Y. (2020). Disclosing the Potential of the SARP-Type Regulator PapR2 for the Activation of Antibiotic Gene Clusters in Streptomycetes. *Front. Microbiol., 11*, 225. https://doi.org/10.3389/fmicb.2020.00225

Laemmli, U. K. (1970). Cleavage of structural proteins during the assembly of the head of bacteriophage T4. *Nature, 227*(5259), 680–685. https://doi.org/10.1038/-227680a0

Lancini, G. & Lorenzetti, R. (1993). Biotechnology of antibiotics and other bioactive microbial metabolites. Language of science. Springer Science+Business Media, LLC.

Laureti, L., Song, L., Huang, S., Corre, C., Leblond, P., Challis, G. L & Aigle, B. (2011). Identification of a bioactive 51-membered macrolide complex by activation of a silent polyketide synthase in *Streptomyces ambofaciens*. *PNAS*, *108*(15), 6258–6263. https://doi.org/10.1073/pnas.1019077108

Laursen, J. B. & Nielsen, J. (2004). Phenazine natural products: biosynthesis, synthetic analogues, and biological activity. *Chem. Rev.*, *104*(3), 1663–1686. https://doi.org/10.1021/cr020473j

Lee, N., Kim, W., Chung, J., Lee, Y., Cho, S., Jang, K.-S., Kim, S. C., Palsson, B. & Cho, B.-K. (2020). Iron competition triggers antibiotic biosynthesis in *Streptomyces coelicolor* during coculture with *Myxococcus xanthus*. *ISME J.*, *14*(5), 1111–1124. https://doi.org/10.1038/s41396-020-0594-6

Lennox, E. S. (1955). Transduction of linked genetic characters of the host by bacteriophage P1. *Virology*, *1*(2), 190–206. https://doi.org/10.1016/0042-6822(55)90016-7

Li, J., Li, Y., Niu, G., Guo, H., Qiu, Y., Lin, Z., Liu, W. & Tan, H. (2018). NosP-Regulated Nosiheptide Production Responds to Both Peptidyl and Small-Molecule Ligands Derived from the Precursor Peptide. *Cell Chem. Biol.*, *25*(2), 143–153.e4. https://doi.org/10.1016/j.chembiol.2017.10.012

Lim, S.-K., Ju, J., Zazopoulos, E., Jiang, H., Seo, J.-W., Chen, Y., Feng, Z., Rajski, S. R., Farnet, C. M. & Shen, B. (2009). iso-Migrastatin, migrastatin, and dorrigocin production in *Streptomyces platensis* NRRL 18993 is governed by a single biosynthetic machinery featuring an acyltransferase-less type I polyketide synthase. *J. Biol. Chem.*, *284*(43), 29746–29756. https://doi.org/10.1074/jbc.M109.046805

Liu, G., Chater, K. F., Chandra, G., Niu, G. & Tan, H. (2013). Molecular regulation of antibiotic biosynthesis in *streptomyces*. *Microbiol. Mol. Bio. Rev.*, *77*(1), 112–143. https://doi.org/10.1128/MMBR.00054-12

Liu, Z., Zhao, Y., Huang, C. & Luo, Y. (2021). Recent Advances in Silent Gene Cluster Activation in *Streptomyces*. *Front. Bioeng. Biotechnol.*, *9*, 632230. https://doi.org/10.3389/fbioe.2021.632230

Locey, K. J. & Lennon, J. T. (2016). Scaling laws predict global microbial diversity. *PNAS*, *113*(21), 5970–5975. https://doi.org/10.1073/pnas.1521291113

MacNeil, D. J., Gewain, K. M., Ruby, C. L., Dezeny, G., Gibbons, P. H. & MacNeil, T. (1992). Analysis of *Streptomyces avermitilis* genes required for avermectin biosynthesis utilizing a novel integration vector. *Gene*, *111*(1), 61–68. https://doi.org/10.1016/0378-1119(92)90603-m

Madhav, N., Oppenheim, B., Gallivan, M., Mulembakani, P., Rubin, E. & Wolfe, N. (2017). *Disease Control Priorities: Improving Health and Reducing Poverty: Pandemics: Risks, Impacts, and Mitigation* (D. T. Jamison, H. Gelband, S. Horton, P. Jha, R. Laxminarayan, C. N. Mock & R. Nugent, Hg.) (3rd). https://doi.org/10.1596/978-1-4648-0527-1_ch17

Martínez-Burgo, Y., Santos-Aberturas, J., Rodríguez-García, A., Barreales, E. G., Tormo, J. R., Truman, A. W., Reyes, F., Aparicio, J. F. & Liras, P. (2019). Activation of Secondary Metabolite Gene Clusters in *Streptomyces clavuligerus* by the PimM Regulator of *Streptomyces natalensis*. *Front. Microbiol.*, *10*, 580. https://doi.org/10.3389/fmicb.2019.00580

Martínez-Hackert, E. & Stock, A. M. (1997). The DNA-binding domain of OmpR: crystal structures of a winged helix transcription factor. *Structure*, *5*(1), 109–124. https://doi.org/10.1016/S0969-2126(97)00170-6

Mast, Y., Guezguez, J., Handel, F. & Schinko, E. (2015). A Complex Signaling Cascade Governs Pristinamycin Biosynthesis in *Streptomyces pristinaespiralis*. *Appl. Environ. Microbiol.*, *81*(19), 6621–6636. https://doi.org/10.1128/AEM.00728-15

Mavrodi, D. V., Bonsall, R. F., Delaney, S. M., Soule, M. J., Phillips, G. & Thomashow, L. S. (2001). Functional analysis of genes for biosynthesis of pyocyanin and phenazine-1-carboxamide from *Pseudomonas aeruginosa* PAO1. *J. Bacteriol.*, *183*(21), 6454–6465. https://doi.org/10.1128/JB.183.21.6454-6465.2001

Misaki, Y., Nindita, Y., Fujita, K., Fauzi, A. A. & Arakawa, K. (2022). Overexpression of SRO_3163, a homolog of *Streptomyces* antibiotic regulatory protein, induces the production of novel cyclohexene-containing enamide in *Streptomyces rochei*. *Biosci. Biotechnol. Biochem.*, *86*(2), 177–184. https://doi.org/10.1093/bbb/-zbab206

Muth, G. (2018). The pSG5-based thermosensitive vector family for genome editing and gene expression in actinomycetes. *Appl. Microbiol. Biotechnol.*, *102*(21), 9067–9080. https://doi.org/10.1007/s00253-018-9334-5

Nayfach, S., Roux, S., Seshadri, R., Udwary, D., Varghese, N., Schulz, F., Wu, D., Paez-Espino, D., Chen, I.-M., Huntemann, M., Palaniappan, K., Ladau, J., Mukherjee, S., Reddy, T. B. K., Nielsen, T., Kirton, E., Faria, J. P., Edirisinghe, J. N., Henry, C. S., . . . Eloe-Fadrosh, E. A. (2021). A genomic catalog of Earth's microbiomes. *Nat. Biotechnol.*, *39*(4), 499–509. https://doi.org/10.1038/s41587-020-0718-6

Newman, D. J. & Cragg, G. M. (2012). Natural Products As Sources of New Drugs over the 30 Years from 1981 to 2010. *J. Nat. Prod.*, *75*(3), 311–335. https://doi.org/10.1021/np200906s

Okada, B. K. & Seyedsayamdost, M. R. (2017). Antibiotic dialogues: induction of silent biosynthetic gene clusters by exogenous small molecules. *FEMS Microbiol. Rev.*, *41*(1), 19–33. https://doi.org/10.1093/femsre/fuw035

Okada, B. K., Wu, Y., Mao, D., Bushin, L. B. & Seyedsayamdost, M. R. (2016). Mapping the Trimethoprim-Induced Secondary Metabolome of *Burkholderia thailandensis*. *ACS Chem. Biol.*, *11*(8), 2124–2130. https://doi.org/10.1021/acschembio.6b00447

Okanishi, M., Suzuki, K. & Umezawa, H. (1974). Formation and reversion of Streptomycete protoplasts: cultural condition and morphological study. *J. Gen. Microbiol.*, *80*(2), 389–400. https://doi.org/10.1099/00221287-80-2-389

Omura, S., Ikeda, H., Ishikawa, J., Hanamoto, A., Takahashi, C., Shinose, M., Takahashi, Y., Horikawa, H., Nakazawa, H., Osonoe, T., Kikuchi, H., Shiba, T., Sakaki, Y. & Hattori, M. (2001). Genome sequence of an industrial microorganism *Streptomyces avermitilis*: deducing the ability of producing secondary metabolites. *PNAS*, *98*(21), 12215–12220. https://doi.org/10.1073/pnas.211433198

Paget, M. S., Chamberlin, L., Atrih, A., Foster, S. J. & Buttner, M. J. (1999). Evidence that the extracytoplasmic function sigma factor sigmaE is required for normal cell wall structure in *Streptomyces coelicolor* A3(2). *J. Bacteriol.*, *181*(1), 204–211. https://doi.org/10.1128/jb.181.1.204-211.1999

Pan, Y., Wang, L., He, X., Tian, Y., Liu, G. & Tan, H. (2011). SabR enhances nikkomycin production via regulating the transcriptional level of sanG, a pathway-specific regulatory gene in *Streptomyces ansochromogenes*. *BMC Microbiol.*, *11*, 164. https://doi.org/10.1186/1471-2180-11-164

Riedel, S., Morse, S. A., Mietzner, T. & Miller, S. (2019). *Medical Microbiology*. McGraw-Hill Education.

Russell, A. D. (2004). Types of Antibiotics and Synthetic Antimicrobial Agents. In S. P. Denyer, N. A. Hodges & S. P. Gorman (Hrsg.), Hugo and Russell's Pharmaceutical Microbiology (S. 152–186). Blackwell Science Ltd. https://doi.org/10.1002/-978047098 8329.ch10

Schneider, C. A., Rasband, W. S. & Eliceiri, K. W. (2012). NIH Image to ImageJ: 25 years of image analysis. *Nat. Methods*, *9*(7), 671–675. https://doi.org/10.1038/nmeth.2089

Schwalbe, R., Steele-Moore, L. & Goodwin, A. C. (2007). *Antimicrobial susceptibility testing protocols*. Taylor & Francis.

Sheldon, P. J., Busarow, S. B. & Hutchinson, C. R. (2002). Mapping the DNA-binding domain and target sequences of the *Streptomyces peucetius* daunorubicin biosynthesis regulatory protein, DnrI. *Mol. Microbiol.*, *44*(2), 449–460. https://doi.org/10.1046/j.1365-2958.2002.02886.x

Sidda, J. D., Song, L., Poon, V., Al-Bassam, M., Lazos, O., Buttner, M. J., Challis, G. L. & Corre, C. (2014). Discovery of a family of γ-aminobutyrate ureas via rational derepression of a silent bacterial gene cluster. *Chem. Sci.*, *5*(1), 86–89. https://doi.org/10.1039/C3SC52536H

Singh, S. B., Genilloud, O. & Peláez, F. (2010). Terrestrial Microorganisms – Filamentous Bacteria. In *Comprehensive Natural Products II* (S. 109–140). Elsevier. https://doi.org/10.1016/B978-008045382-8.00036-8

Singh, S. B., Jayasuriya, H., Ondeyka, J. G., Herath, K. B., Zhang, C., Zink, D. L., Tsou, N. N., Ball, R. G., Basilio, A., Genilloud, O., Diez, M. T., Vicente, F., Pelaez, F., Young, K. & Wang, J. (2006). Isolation, structure, and absolute stereochemistry of platensimycin, a broad spectrum antibiotic discovered using an antisense differential sensitivity strategy. *J. Am. Chem. Soc.*, *128*(36), 11916–11920. https://doi.org/10.1021/ja062232p

Soltero., F. V. & Johnson., M. J. (1953). The effect of the carbohydrate nutrition on penicillin production by *Penicillium chrysogenum* Q-176. *Appl. Microbiol.*, *1*(1), 52–57. https://doi.org/10.1128/am.1.1.52-57.1953

Takano, E., White, J., Thompson, C. J. & Bibb, M. J. (1995). Construction of thiostrepton-inducible, high-copy-number expression vectors for use in *Streptomyces* spp. *Gene*, *166*(1), 133–137. https://doi.org/10.1016/0378-1119(95)00545-2

Tanaka, A., Takano, Y., Ohnishi, Y. & Horinouchi, S. (2007). AfsR recruits RNA polymerase to the *afsS* promoter: a model for transcriptional activation by SARPs. *J. Mol. Biol.*, *369*(2), 322–333. https://doi.org/10.1016/j.jmb.2007.02.096

Tanaka, Y. & Omura, S. (1990). Metabolism and Products of Actinomycetes—An Introduction. *Actinomycetologica*(4), Artikel 1, 13–14. https://doi.org/10.3209/saj.4_-13

Thompson, C. J., Ward, J. M. & Hopwood, D. A. (1982). Cloning of antibiotic resistance and nutritional genes in streptomycetes. *J. Bacteriol.*, *151*(2), 668–677. https://doi.org/10.1128/jb.151.2.668-677.1982

Tsuji, N., Kobayashi, M., Wakisaka, Y., Kawamura, Y. & Mayama, M. (1976). New antibiotics, griseusins A and B. Isolation and characterization. *J. Antibiot.*, *29*(1), 7–9. https://doi.org/10.7164/antibiotics.29.7

Waksman, S. A. & Fenner., F. (1949). Drugs of natural origin. *Ann. NY Acad. Sci.*, *52*(5), 750–787. https://doi.org/10.1111/j.1749-6632.1949.tb53965.x

Waksman, S. A. & Woodruff, H. B. (1940). Bacteriostatic and Bactericidal Substances Produced by a Soil Actinomyces. *Exp. Biol. Med.*, *45*(2), 609–614. https://doi.org/10.3181/00379727-45-11768

Wex, K. W., Saur, J. S., Handel, F., Ortlieb, N., Mokeev, V., Kulik, A., Niedermeyer, T. H. J., Mast, Y., Grond, S., Berscheid, A. & Brötz-Oesterhelt, H. (2021). Bioreporters for direct mode of action-informed screening of antibiotic producer strains. *Cell Chem. Biol.*, *28*(8), 1242–1252.e4. https://doi.org/10.1016/j.chembiol.2021.02.022

Wietzorrek, A. & Bibb, M. (1997). A novel family of proteins that regulates antibiotic production in streptomycetes appears to contain an OmpR-like DNA-binding fold. *Mol. Microbiol.*, *25*(6), 1181–1184. https://doi.org/10.1046/j.1365-2958.1997.5421903.x

Wohlleben, W., Bera, A., Mast, Y. & Stegmann, E. (Hrsg.). (2017). Chapter 8—Regulation of Secondary Metabolites of Actinobacteria. https://doi.org/10.1007/978-3-319-60339-1_8

Ye, J., Zhu, Y., Hou, B., Wu, H. & Zhang, H. (2019). Characterization of the bagremycin biosynthetic gene cluster in *Streptomyces* sp. Tü 4128. *Biosci. Biotechnol. Biochem.*, *83*(3), 482–489. https://doi.org/10.1080/09168451.2018.1553605

Yu, T. W., Bibb, M. J., Revill, W. P. & Hopwood, D. A. (1994). Cloning, sequencing, and analysis of the griseusin polyketide synthase gene cluster from *Streptomyces griseus*. *J. Bacteriol.*, *176*(9), 2627–2634. https://doi.org/10.1128/jb.176.9.2627-2634.1994

Printed in the United States
by Baker & Taylor Publisher Services